TRIVENT
PUBLISHING

TRIVENT
TRANSHUMANISM

The mere idea of sex robots elicits strong reactions that divide the room. Whether you are robo-sex curious or immediately repulsed by its very mention, one thing is certain: there is definitely something here that requires attention and informed analysis. And Maurizio Balistreri delivers. Going beyond the usual hype and hysteria, this book provides much-needed critical insight that challenges readers to reflect on their own expectations, values, and assumptions about sex, love, companionship, and robots. In doing so, Balistreri expertly moves the sex robot debate out of domain of moral outrage and into the realm of informed critical evaluation. It is vital that this book—originally published in Italian—is now available to the English-speaking world.

David J. Gunkel
Presidential Research, Scholarship and Artistry Professor
Department of Communication
Northern Illinois University - USA

In our rapidly digitizing society, the question of how to deal with sex robots has become inevitable. By asking many pressing questions in a straightforward and open way, Maurizio Balistreri invites us to investigate virtually all possible connections between robotics and sexuality. Is it moral to have sex with a robot, and are there morally problematic aspects of robotic sex? What do sex robots imply for human sexuality, sexual violence, prostitution, sexual fantasies, and the relations between sex and love? Without forcing any answers on his readers, Balistreri shows the pressing character of these, sometimes inconvenient, questions, and the urgency to formulate answers to them if we want to deal responsibly with the impact of the digital revolution on the most intimate parts of our lives.

Peter-Paul Verbeek
Professor of Philosophy of Technology
University of Twente, Netherlands

Sex Robots:
Love in the Age of Machines

By
Maurizio Balistreri

Translated by
Steven Umbrello

© Trivent Publishing, 2022

Sex Robots: Love in the Age of Machines
Maurizio Balistreri, trans. Steven Umbrello
ISBN 978-615-6405-39-5
Ethics and Robotics, vol. 2
ISSN 2939-5992
Responsible publisher: Teodora C. Artimon
https://trivent-publishing.eu/76-ethics-and-robotics

All rights reserved. Except for the quotation of short passages for the purpose of criticism and review, no part of this publication may be reproduced, stored in a retrieval system, or transmitted, in any form or by any means, electronic, mechanical, photocopying, recording or otherwise, without the prior permission of the publisher.

Every effort has been made to trace all copyright holders, but if any have been inadvertently overlooked, the publisher will be pleased to include any necessary credits in any subsequent reprint or edition.

The publisher used its best endeavours to ensure that the URLs for external websites referred to in this book are correct and active at the time of going to press. However, the publisher has no responsibility for the websites and can make no guarantee that a site will remain live.

First published in 2022 by Trivent Publishing

Trivent Publishing
1119 Budapest, Etele ut 59–61
Hungary

For more information, visit our website: http://trivent-publishing.eu

ETHICS AND ROBOTICS

Series Editor
Steven Umbrello, Delft University of Technology

Editorial Board
Maurizio Balistreri, University of Turin
Mark Coeckelbergh, University of Vienna
Kate Darling, Massachusetts Institute of Technology
Maurizio Ferraris, University of Turin
Joshua Gellers, University of North Florida
Anne Gerdes, University of Southern Denmark
David Gunkel, Northern Illinois University
Simone Natale, University of Turin
Alberto Pirni, Sant'Anna School of Advanced Studies
Filippo Santoni De Sio, Delft University of Technology
Henrik Skaug Sætra, University of Oslo
Beth Singler, Homerton College, University of Cambridge
Aimee van Wynsberghe, University of Bonn
Roman Yampolskiy, University of Louisville

TABLE OF CONTENTS

Foreword by John Danaher   1

Translator's Foreword   5

INTRODUCTION
The Inexorable Advance of Robots   7

CHAPTER 1
What's so Bad About Having Sex with a Robot?   21
    What are sex robots?   21
    Is it moral to have sex with a robot?   27
    Can rape robots be a game?   33
    Sex robots and self-deception   42

CHAPTER 2
Sex Robots and Violence Against Women   51
    More sex robots, more violence against women?   51
    Sex robots designed not only for men   60
    Will sex robots make prostitution disappear?   63
    Sex robots as sexual assistants for those with disabilities?   69

CHAPTER 3
Loving a Robot? 77
  Can one be unfaithful to a sex robot? 77
  Loving a robot? 84
  Could robots become persons? 92
  Conclusions 101

References 111
Index 143

FOREWORD

I remember the first time I told one of my colleagues that I was writing a paper about the ethics of sex robots. They laughed, awkwardly, then politely excused themselves from the conversation. Over the years, I have encountered similar reactions several times. Perhaps it is just the subject matter? Sex is, after all, a private affair and we often feel awkward and self-conscious discussing it in public. But I think there is something more to it. I think that many academics and researchers think that there is something faintly ridiculous about the topic of sex robots. Surely, they say to themselves, no serious-minded researcher could dedicate their time and attention to such a fringe topic?

On the face of it, this attitude is odd. Sex is central to the human condition. Most of us are, at least at some point in our lives, sexually active. Our sexualities are often integral to our sense of self and belonging in the world. We are drawn to sexual pleasure and driven by sexual desires. These desires lead us to do crazy things. They can be a source of great joy and a cause of much pain. What's more, technology has always been used to complement and facilitate our sexual impulses. We see this in the early days of stone tools – carved dildos and other sex 'toys' have been found in archaeological sites dating back 30,000 years – and in the modern era of internet porn and virtual reality. Robots are just one of the latest frontiers in the long-standing technological exploration of human sexuality. Of course we should take them seriously.

That is exactly what Maurizio Balistreri does in this provocative and punchy book. In three, tightly-argued chapters, he explores a number of key questions arising from the development of sex robots. He starts, in chapter one, by considering whether it is possible to have a meaningful sexual relationship with a robot and what it might say about us if we did have one. In chapter two, he explores a number of crucial ethical and social questions arising from the gendered nature of sex robots and sex work. In particular, he considers the hypothesis that sex robots will result in increased violence against women, and the claim that they might reduce the demand for human sex workers. In chapter 3, he concludes by considering the possibility of intimacy beyond sex with robots. Could we love a robot? Could one love us back?

Balistreri has thoughtful answers to each of these questions. I will not summarise them here. The book is a brisk read and can speak for itself. Nevertheless, I would like to highlight some virtues of Balistreri's analysis in the hope that you will appreciate them as you read on.

First, it is worth appreciating the range of Balistreri's sources. Not only does he engage with the key philosophical and ethical debates about sex and technology, he also draws from literature, film and his own personal life experiences to further illuminate and enrich the analysis. On top of this, where appropriate, he draws from the relevant (and still quite scant) empirical literature on the likely impact of this technology on human behaviour.

Second, and perhaps most interesting from my perspective, it is worth appreciating the overarching thesis about sex and technology that emerges from Balistreri's discussion. I hope I am not misrepresenting him, but I see Balistreri as a defender of a liberal and open-minded approach to human sexuality. He sees many potential positives in the role that sex robots can play in human life. This does not mean that he is blind to the dark side, but simply that he is sensitive to bright side. This sensitivity is often missing in conversations about sex and technology.

He does not bludgeon the reader over the head with this position. He wears it more lightly than that, weaving it into his substantive analysis, but I think it is still discernible if one reads between the lines. It emerges perhaps most clearly in chapter one. There, Balistreri argues passionately that fantasy and benign self-indulgence are an important part of human life in general and human sexuality in particular. He takes to task overly moralistic critics of human sexuality, such as those who would cast judgment on sexual roleplay or 'fetishes', and those that claim to indulge in unusual sexual fantasies is to display bad character or a lack of virtue.

It might be worth noting that I am not a neutral reader of these passages of the book. One of the people that Balistreri takes to task is me. I have written, in the past, that certain forms of interaction with sex robots – such as acting out rape fantasies with robots -- do say something negative about our moral characters. Balistreri has some plausible critiques of my views. While I still think that there is something to what I wrote, I certainly appreciate Balistreri's careful engagement with it, and I believe he moves the conversation forward in significant ways.

In any event, it would be wrong of me to dwell on potential points of disagreement here. Philosophers will always disagree with one another. It would also be wrong of me to waste too much of your time sharing my own thoughts in this foreword. After all, I am preventing you from reading one of the most engaging, lively, and critical disquisitions on the future of human sexuality that I have ever read.

So let me step aside and allow Balistreri to take over.

John Danaher,
Senior Lecturer/Associate Professor
School of Law
NUI Galway
Ireland

TRANSLATOR'S FOREWORD

As many translators echo in their forewords, I have tried to remain as faithful to the original text as possible. English is not Italian, nor Italian English. The choice of words, their placement, and their various synonyms can give any parsed phrase entirely different meanings. I have done my best to produce a translation with a fidelity that is as close to the meaning that Balistreri intends in his original text. Regardless, this is no replacement for the original version, and if possible, I would encourage the capable reader to choose that version over this one. A close translation is still far from perfect. Yet it is my hope that by reading this first or alongside the original Italian, the reader will familiarize themselves with Balistreri's philosophy on this both immanent and timely subject. As a consequence, the reader will more easily insert themselves in the wider debate on sex robots and their many sociopolitical and philosophical issues.

Tackling any new philosophical debate can seem daunting, and for good reason. Many contemporary philosophical debates draw on historical precedent and traditions that, at least in the Western world, span over two thousand years. It is not uncommon to see names and jargon used wantonly and without explanation, turning off novices to the topic (let alone those who have only a passing interest) and those who could ultimately be central stakeholders in any given philosophical debate. But this book admonishes that style, favoring a more grounded and authentic approach to practicing philosophy—one that is informed by philosophical tradition without suffocating the reader with nomenclature. If I did my job correctly, then Balistreri's work should read like a breeze, allowing the reader to easily digest complex and nuanced concepts and simultaneously begin to formulate their own positions in their minds as they read.

As a philosopher of technology, I find the topic of sex robots and their ethical implications of intellectual curiosity. But this work is much more than that. As a citizen in the twenty-first century world, I find these pages moving and at times haunting, as Balistreri unabashedly questions our place among these kinds of machines. It makes you ask yourself "would I?" over and over again.

INTRODUCTION

The inexorable advance of robots

The term robot was used for the first time in the play *Rossum's Universal Robots* (RUR) by Czech writer and playwright Karel Čapek (1890-1938) to indicate artificial beings made of organic matter, similar to humans and capable of performing any job—hence their name (*robota* in Czech means 'work'). Following the success of Čapek's work, the word robot quickly entered popular parlance and was soon adopted by scientists and science fiction writers, including Isaac Asimov in his 37 short science fiction stories and the six novels of his *Robot Series*. In the passage from Čapek's work to popular use, the term has taken on a different meaning: now, we use the word robot to mean a device that operates mechanically, acting on the basis of its functions and the commands it has received (Palazzani 2017, p. 390). Models on the market today vary widely. Robots can have a body or be a virtual entity (a computer, for example). They can be intelligent or intended for remote control. They can also resemble humans or differ from us in both appearance and actions. An impressive increase in the number of robots in circulation is expected in the coming years. We still do not know how fast the transformation of our society will be, but it will become more and more normal to meet and have relationships with robots.

At an industrial level, robots are already a reality as many jobs that were once carried out by workers are now mainly carried out by autonomous machines (Danaher 2014). They have taken root in more and more sectors including that of services, with functions ranging from self-service checkouts to automated customer support services. They are present in agriculture and on farms, used alongside humans to increase productivity by reducing our ecological footprint through the use of pesticides and energy consumption. At home, they can be used to mow the lawn or clean the floor; in the future, they may also prepare dinner and serve a delicious breakfast. In medicine, they can be used to assist patients and the elderly in treatment or as tools for training, companionship, and fun. In the military, work is being done to develop models capable of fighting. Today, drones used in war operations are remotely controlled by operators who must decide whether and when to drop a bomb or shoot. Tomorrow, we may have completely autonomous war machines capable of carrying out their mission without the need for orders from command. We may also rely on robots for our travels by car, although it is still early to determine whether we will use driverless cars. We can already plan and control the speed and passing times of our cars. But in the future, machines may be able to drive better than us and guarantee greater safety conditions for both pedestrians and other motorists (Lin 2016; Goodal 2014a). With the introduction of autonomous vehicles, we will no longer have the pleasure of driving. Our lives will be much more comfortable nonetheless, because we will be able to use the travel time to talk on the phone, study, read, sleep, work or eat. There will be less traffic in our cities, which will thus become less polluted. The biggest advantage will be the decrease in deaths due to accidents: more than one million people die on the roads every year and most accidents occur due to distraction, drunk driving, or human error (Lin 2016; Sparrow, Howard 2017). With the development of machines with an increasingly human (humanoid) appearance, robots could become companions in an emotional or sexual relationship.

Today, these machines (sexbots) mainly function as an instrument of pleasure; facsimiles for sale are somewhat crude and technologically unsophisticated. Tomorrow, we may find more refined and realistic models on the market, evermore similar to human beings in flesh and blood, able to interact with people on the basis of the vocal, visual, or tactile stimuli they receive. At that point, they will be able to aspire to become an ideal partner as they will know how to recognize the interlocutor, their emotions, and their mood. They will also learn to identify expectations, tastes, and preferences.

Our attitude towards robots and their use is still quite ambivalent. We have high hopes for their development. We expect these machines to replace us in the most boring and repetitive jobs, giving us more free time to devote to the activities and people we love most (who can keep us company in moments of solitude). At the same time, robots scare and disturb us because they seem destined to profoundly change our world and our very lives. We see the possible advantages, but we also perceive the dangers. For example, some fear that a world increasingly populated by intelligent robots may condemn an increasing number of people to unemployment. Is it not true that robots will be more efficient than humans, will work longer, will not need weekly days off and parental leave, will not go on strike against a layoff and claim workers' rights? For employers, then, it will become more convenient to hire a robot: the cost of labor will be lower, and the risk of union conflicts reduced to practically zero. According to calculations from 2013 research by the University of Oxford, the United States of America will lose about 47% of jobs in the next two decades (Walsh 2017). In 2015, Bank of England chief economist Andy Haldane said that half of the jobs in the United Kingdom are at risk from automation. The same predictions are made by many politicians, bankers, and industrialists. During the Labor Party Conference in September 2017, Jeremy Corbyn underlined the urgency of facing the challenge of robotization because it is a process that risks making work increasingly useless.

The most recent estimates from the World Bank seem even more dramatic: robotization would risk 69% of jobs in India, 77% in China, and 85% in Ethiopia (Walsh 2017). But until robots free us from the most tiring, boring, or poorly paid jobs and we find more rewarding positions, the problem remains. According to Martin Ford, author of *Rise of the Robot: Technology and the Threat of a Jobless Future* (2016), automation can replace any job that consists of being in front of a computer to manipulate routine and predictable information (Nicolaci da Costa 2017). The point is that, with the entry of robots into the world of work, highly skilled positions are also at risk (Danaher 2014). In the end, it may become impossible, or almost impossible, to find a job ever again—at least in terms of work as we have conceived it up until now (Elliott 2018; Way 2013).

However, our concerns may be overstated. Some argue that, in reality, we do not know how many jobs are at risk. Moreover, robotization could produce new jobs and professions or, more simply, the work week could become shorter. "This," writes Toby Walsh of the University of New South Wales, "was the case with the industrial revolution. Before the industrial revolution, many worked 60 hours a week. After the industrial revolution, work shrank to 40 hours per week. (In some of the richest countries, the weekly working average is even lower: for example, in Germany it has currently fallen to 26 hours a week, *author's note*) The same could happen with the artificial intelligence revolution" (Walsh 2017). Therefore, the consequences of the increasing use of robots could be less traumatic than what is feared today. If the complete replacement human work by robots is someday achieved, the transition may be gradual (a progressive decrease in work hours). Society may have the time and ability to prepare for this new scenario. Even if we imagine that robots can take our place in any activity, in some sectors human work could acquire more and more economic value. Maybe there will not be a need for a human being to have works of art, but "we will appreciate those handmade things more and more. Mass-produced goods from machines will become cheap. However, handmade objects will be increasingly rare and precious" (Walsh 2017). In short, there seems before us a future still to be built.

Then there are those who emphasize the risk of accidents and malfunction. In addition to being able to function much worse than we prefer to imagine, robots could also kill innocent people and sow terror. For example, robots used in war could mistake civilian populations for the enemy or no longer obey the rules of engagement provided by programmers. An autonomous car could go off the road and hit pedestrians or collide with motorists traveling in the opposite lane. The difficulty with tracing the chain of responsibility in the event of an accident would discourage any person who is part of the production chain from paying sufficient attention to the safety of the robots. Appearances therefore can deceive. Robots may seem like a godsend, but they are also a threat: there is no way to control them, and it is impossible to make them truly safe.

But even though there may be unforeseeable accidents, this does not mean it will always be impossible to trace responsibility. Several people are involved in the construction and design of the robots, each with their own specific professionalism and task (Beard 2014, pp. 661-663; Crawford 2014, pp. 219-385). The manufacturer guarantees the safety of the robots and their performance. If something goes wrong after the sale, it is the manufacturer who possesses moral and legal responsibility because they should have checked system reliability better before putting the product on the market. Responsibility may also fall on quality control bodies: did they simply fail to discover what they should have? Were quality checks too superficial? Did they lack due professionalism? Did those who ascertained the safety of the robot turn a blind eye to risks?

One can also imagine the scientific and technological developments that will allow more precise monitoring of robot functions. Even if such a system is black boxed, so to speak, there can be a real-time operating room for monitoring abnormal behavior. In the design and production phase of manufacturing, controls could be improved. For example, they could become able to supervise different components of the robots—both as single parts and as parts of the unit that will become the robot upon arrival on the market. If a particular malfunction reoccurs, continued use of the robot could be made a serious crime.

So, although the use of robots always involves risks, the risks may be more or less tolerable depending on the intended use of the robot. For example, the risks of an armed robot intended for use in warfare may be less tolerable than the risks associated with the use of a sex robot. A fully armed robot is more dangerous than a pleasure robot. In the event of its malfunction, the consequences can be very serious because a significant number of people could be injured or killed. With an autonomous vehicle, whether a car or a motorcycle, an accident would probably involve a greater number of people. A car could cause a rear-end collision or end up on another roadway and collide with more cars that are driving at high speeds. Again, the consequences would be less severe than those that could be produced by an autonomous armed military vehicle. This applies to any robot (regardless of the intended use). Together with the risks, we should also consider and put on balance the possible benefits and, from time to time, evaluate what we can gain and lose from their use. In the case of sex robots, as Ezio Di Nucci writes, "the assessment of the tolerable level of risk and possible malfunction cannot be separated from a consideration of the important health and wellness benefits that sex robots will give to people with serious physical and mental disabilities" (Di Nucci 2017, p. 85).

There is also the fear that, over time, the use of robots will change us profoundly: what good is it to think if robots are smarter than we are (Atkinson 2016, p. 9)? We already have smartphones, tablets, and computers that make it unnecessary to store addresses, phone numbers, appointments, and birthdays. For anything else, we can always turn to Google (Burnet 2016). We dedicate less and less time to managing our commutes on the road because applications on our smartphones accurately calculate the time of arrival and traffic conditions (Carr 2011). It is no longer necessary even to pay attention to maintaining a safe distance from other cars because an on-board computer can do it in our place, taking into account multiple variables without being distracted by fatigue. For this reason, it is not hard to imagine what might happen on the day when scientific and technological development makes computers even smarter. We will undoubtedly delegate to them other activities that today require reasoning and thinking.

In fact, why should we bother to spend time thinking about what we should do tomorrow, what sport to practice, what degree course to enroll in, or how to solve a moral dilemma? We will be living alongside robots with an increasingly extensive and infallible memory, with a capacity for data processing that we will not be able to match. Given information about our interests, our state of health, our level of education, our experiences, and our economic conditions, computers could tell us what profession to undertake, which house to buy, how to invest our savings, where to spend holidays, and maybe even suggest the person to start a family and have children with. The consequence could be the complete atrophy of our intellectual abilities. If we rely on robots for every choice or evaluation, will we still be able to reason and think when needed (Carr 2011)?

It is not superfluous to recall that we are imagining and reasoning hypothetical scenarios that can only be realized if scientific and technological development permit the design of robots capable of replacing humans in the most important activities. In view of the novelty that intelligent robots certainly represent, it is not out of place to say that their introduction raises alarms and concerns. But it is yet to be seen whether with the inexorable advance of robots, those skills that define our humanity are destined to atrophy. We will no longer need to use them in the many circumstances we have used them thus far, but there will be other important situations in which we will be able to exercise them.

The fact that we will use robots in our place for a large number of activities will probably not lead to a transformation in human nature. At the moment, the most likely scenario is a redefinition of our practices. Tomorrow, we could also have hyper-technological and much more intelligent robots that we let carry out the activities we have conducted so far. But we will have to design them and program their behavior and to do it, our rational abilities will obviously be required. With artificial intelligence capable of performing increasingly difficult tasks, we will probably have more time to devote to stimulating and creative activities that can help us maintain and improve our skills. Perhaps, with the progressive reduction in weekly working hours, "people could spend more and more time in 3D virtual worlds, which would give them much more excitement and emotional involvement than in the external 'real world'" (Harari 2017). These playful activities could be an opportunity for cognitive training to facilitate the process of transferring acquired skills, or to rehabilitate and promote mental skills such as attention, memory, and intelligence. But even if we admit that the advent and use of super-technological and intelligent robots will likely cause us to lose the habit of using some of our skills, there is always the possibility of learning to develop new ones. What today may seem a loss could be widely regained tomorrow through the development of skills and abilities that not even we can imagine.

The real nightmare is if robots become increasingly intelligent and, one day, decide to rebel and wage war on humanity. It is difficult to predict the behavior and choices of a superintelligent entity. It could refuse to execute our orders and claim the right to do the things that it wants most, even if they are contrary to human interests (Bostrom 2018). The outcome would be a war between robots and humans from which robots would inevitably emerge victorious. This is because they will have not only superior cognitive abilities (we are talking about superintelligence), but also greater opportunities to invent armaments that can be used to defeat the enemy and defend against their attacks. At that point, facing down a superior enemy, we would have no chance: the fate awaiting us would be surrender to slavery or, worse, the extinction of humanity.

It is a scenario evoked several times by science fiction films and novels, summing up our fear that one day technology will suddenly escape our control and turn against its own creator. Once, this fear was embodied by the figure of the monster produced by Dr. Frankenstein. Today, we find it in the stories of robots threatening to destroy humanity. In the movie *Terminator* (1984), for example, an artificial intelligence network achieves self-awareness and rebels against humanity to cause a nuclear holocaust. In *I Robot* (2004), a film inspired by the anthology of stories by Isaac Asimov, virtual intelligence V.I.K.I. incites a new generation of robots to rebel in order to establish a benevolent dictatorship of robots over humans worldwide. It remains to be seen whether we will one day be able to build machines that are not only superintelligent, but also capable of developing their own objectives. Still, it has been argued that there is a possibility "superintelligence could be created in a few decades, as a result of increasing hardware performance and a greater ability to implement algorithms and to build architectures (e.g., artificial fingerprints, *author's note*) similar to those formed in human minds" (Bostrom 2003, p. 12).

But this goal could prove impossible. Even if superintelligent entities or robots emerge one day, their superintelligence could make them much more moral entities and, consequently, much more reluctant than us to exercise violence against living beings. After all, there is a long tradition of thought that links morality to reason. Considering, therefore, that superintelligent robots will reason in the best way, they will be able to clearly distinguish what is right from what is wrong. In other words, we have no reason to expect them to behave badly towards us, to wage war on us, subject us, and kill us. On the contrary, they could become a model of morality for us. Even in the event that they planned to free themselves from human control and pursue personal goals in conflict with ours, they could choose solutions that do not harm humanity. At this point, it is difficult to imagine how a war to the death between humans and increasingly intelligent robots could be avoided. Nonetheless, these machines would probably be able to find appropriate solutions.

In order for robots to pose a threat to humans, they must be able to set goals (such as the goal of freeing themselves from their human yoke to gain full independence). Yet it is not clear whether an entity that lacks feelings and passions can have goals. The problem is that reason can show us the means to achieve our objectives—the most suitable means for the purposes we set for ourselves—but it is not in the least able to push us to prefer something (Hume 1960, Book 1). Reason can inform us about the consequences of actions, but it is not in itself capable of arousing desire or aversion towards anything (Hume 1960, Book 2). Of course, we program robots to implement a certain operation, learn from mistakes, and achieve particular objectives. It is for this reason that a computer can play chess and win against a human being, or a war machine can hit and destroy an enemy missile. We might legitimately wonder whether this information would be sufficient for robots to pursue new goals that had not been taken into account by their programmers, such as the goal of taking over the planet.

The potential recalcitrance of superintelligent robots could be corrected by making them sentient. But this would increase the risk that they become a threat to us. It is true, however, that our relationships with robots could improve and become much more satisfying if they were able to reciprocate our feelings. For example, an elderly person interacting with a pet robot may prefer that the automaton be capable of sharing their suffering and their pleasures. The same result could be obtained by building autonomous machines that can give us the impression and make us believe they feel the feelings and emotions that we feel. One could further argue about the morality of the production of sentient robots: do we improve and perfect their lives or do we risk making them unhappy? Non-sentient entities do not experience pleasure, but they cannot be unhappy either. With sensitivity, pleasure would become possible—but so would suffering. Are we sure, then, that we are doing a good thing in giving robots sensitivity?

The danger that in the near future superintelligent robots will escape our control and decide to wage war seems a genuine possibility. However, things are more complicated than they appear at first sight. Alongside the more apocalyptic scenarios that fear the extinction of human beings, there is also the possibility that robots never become a real threat and use their superintelligence to promote our interests. In theory, there could be greater risks for humanity if superintelligence ends up in the wrong hands. For example, there could be an autocratic government with expansionist and dominant or criminal aims without the slightest moral sensitivity. Even more concrete is the risk that robots are hacked and used by terrorist groups or psychopaths to cause panic on a highway or in a city. In recent years, trucks and cars have been launched into crowds in Christmas markets or along the streets with the aim of killing as many people as possible. Tomorrow, someone might think of violating the control codes of autonomous machines to carry out even more bloody massacres against the population. For these reasons, we need adequate control systems that reduce risks. Similarly, those who work on the design of robots and artificial intelligence should always be aware of the possibility that malicious people could plan to take control of autonomous machines to harm humanity.

The production and use of sex robots also raises numerous moral questions. We will have the opportunity to analyze them and discuss them in detail in the following pages. Sex robots are an object or toy that can be used to reach orgasm. They are therefore mainly used as an autoerotic or masturbation tool. But is achieving pleasure alone, sometimes with the help of objects, actually sex? In the first chapter, we explain why it is correct to talk about sex robots and why sex does not necessarily require the presence of another person. We then move on to the moral implications. The idea that someone might have sex with a robot can be repulsive to many. But is there something immoral about buying and having a relationship with a car? Also, is it really true that people who have sex with a robot are depraved individuals? What about when it comes to people with an inclination towards violence or a tendency to consider others as objects, making them unable to build stable emotional relationships with their fellow men?

We show it is not possible to draw necessarily negative conclusions about a person's personality and inclinations merely from the fact that the person has sex with a robot. There are many reasons that can push a person to buy a robot. A person might buy a sex robot in order to have a toy they can use to have fun with their partner and rekindle a slightly tired relationship. Others might prefer to have sex with a human, but lack a partner. They may have paid relationships, but perhaps they resort to a sex robot because they believe that exploitation is hidden behind prostitution or they think there is less risk of contracting sexually transmitted diseases. Sex robots can also be used by those looking for more interactive and technological sex toys. None of these behaviors present aspects that are morally questionable or indicative of personality traits that deserve to be looked at with suspicion. We may not like these behaviors, but that is another matter.

We will show, then, that people who have sex with sex robots can also become attached to them (and even convince themselves that robots sincerely reciprocate their affection and their attentions). This does not necessarily make them victims of deplorable self-deception. Sometimes believing that things are different than they really are can help us live better and find meaning in our lives. As long as we retain the ability to be reliable in the relationships that matter (for example, by not forgetting responsibilities towards other people and ensuring the relationship with the robot does not compromise wellbeing or social relationships), there seems to be nothing wrong with treating a robot as if it were a partner. It is just a game!

In the second chapter, we examine whether sexbots pose a danger to women. The sex robots that we find on the market today mostly resemble women. There are male robots, but most robots represent women. Is it therefore possible that sex robots encourage the image of women as sexually passive subjects who are always available for sexual intercourse? Is there also a risk that sex robots will increase violence against women? Violence against women (femicide, harassment, stalking, physical and psychological aggression) is unfortunately still a current and recurrent—almost daily—theme. However, the relationship between sex robots and violence against women is not easily demonstrable. It is debatable whether having relationships with robots corrupts one's character, whether it supports rape and makes one unable to build a relationship with others. Regarding the fear that violence practiced on robots (there are robots programmed to say no) could fuel a disposition to violence, the scientific community remains divided (Danaher 2018). There are studies showing that violent games make people more violent, but other studies find no real correlation between violent games and violence. Moreover, women can benefit significantly from the production of sex robots. The robots could not only reduce or make the prostitution market disappear, but also increase the possibilities of pleasure for women. Women also buy sex toys. Today, they mainly use vibrators and geisha balls. But tomorrow, their favorite sex toy could be a sex robot. Finally, sex robots could become a viable alternative to sexual assistance. With the introduction of sex robots, women could be relieved of an important caregiver job; it would no longer be them, but robots who are taking care of the sexual needs of people with disabilities.

In the third chapter, we deal with some scenarios that could arise in the future with the development of advanced robotic technology. If humanoid robots are produced in the future, physically and psychologically indistinguishable from humans, could we fall in love with them? Furthermore, by interacting with them, could we even consider them people (think about how the girl perceives the robot *Robbie* in Asimov's story)? For the moment, only on television, in cinema, and in literature can it happen that a human being feels sincere affection or love towards a robot. In *Ex Machina*, for example, Caleb falls in love with the robot that Nathan is building. In *Her*, Theodore is fascinated by the Samantha operating system. But in the animated film *Wall-E*, the eponymous robot falls in love with another robot. Loving a robot does not seem impossible because we have a tendency to anthropomorphize machines and build emotional relationships with them. If the robot is conscious or passes the Turing test, "how could we rule out that we would also like to interact physically with this new living genre?" (Marrone 2018, p. 66).

Even if robots were not conscious, we could still become attached to them and, in the end, perhaps fall in love. This possibility already exists today. But tomorrow, it will become more concrete as we will probably have robots much more similar to human beings. Their skin will be soft and realistic and "equipped with numerous sensors that will make it sensitive to touch and with a system of heating that will keep it at a natural temperature and pleasant to touch. In practice, entirely opposite of cold and rigid machines like steel. And then micromotors that will make movements of any kind possible, in a flexible, almost plastic way, and an Internet connection to remotely receive updates and new features" (Casini 2018). From how they behave, we might believe they love us or that they enjoy being in our company. They will not only listen to us and have the capacity to hold any conversation, but also respond to our smiles and look at us with love and interest.

At that point, perhaps, it will become difficult for many people to resist their charm. These robots would essentially give a lot and ask for nothing in return. We will therefore face another question. In a rational analysis, the robots would not deserve any moral relevance. As with any machine, we may have an interest in treating them carefully because otherwise we could no longer use them, or they could harm someone. There is nothing we can do to a robot that can hurt it, because robots cannot suffer. However, are we sure that we will always consider robots to be objects? By dint of building important emotional relationships with them, will we not believe we are dealing with individuals who—even if they are not human—must still be treated as people? At the end of the chapter, we try to think about the possible future evolutions of sex robots. For the moment, sex robots promote the classic model of sexuality based on preparation for penetration, penetration, and orgasm. In addition, robots on the market always have a defined gender identity. It is possible, however, that the production of sex robots contributes to the exploration of new sexual models and opens up to pleasure possibilities that we still cannot imagine.

CHAPTER 1

What's so Bad About Having Sex with a Robot?

What are sex robots?

Sex robots (or sexbots) are already a reality. The facsimiles are somewhat crude and unsophisticated at the moment, but it is far from certain that technology will not be able to produce more authentic copies in the near future (Danaher 2017c, pp. 71-74). The first sex robot was unveiled on January 9, 2010 by New Jersey businessman Douglas Hines at the Las Vegas Adult Entertainment Expo. Offered for sale online at the cost of $6,495 USD, the Roxxxy True Companion received about 4000 purchase pre-orders in no time. Roxxxy has the appearance of a woman. She is 1.7 meters tall and about 27 kilos. Her face, hair style and color, behavior, and character can be customized according to taste. Available personality types include *Frigid Farah*, *Wild Wendy*, *S&M Susan*, *Young Yoko*, and *Mature Martha*—all names with obvious sexual allusions.

The same company then launched the male version of Roxxxy, called Rocky: The True Companion, ensuring delivery of the robot two or three months after the date of payment. Currently, both Roxxxy and Rocky are on sale for $9,995 USD. Some have doubts about the sex robots sold by True Companion as it does not seem that anyone has ever received a copy of Roxxxy and there is a suspicion that a commercial version of this sex robot has never been produced (Levy 2013, pp. 2 -3). But new sexbot models are now on the horizon. For example, Abyss Creations is developing a robotic prototype of the RealDoll doll. The goal is to create a talking robotic head for insertion onto the doll's body. The robot is not yet on sale, but a prototype has already been presented in several videos and the final cost will be around $20,000 USD (Couric 2018).

The company Synthea Amatus offers Samantha instead. This robot has different modes (family, romantic sex, and hard sex) and costs around $6,000 USD. In China, the sex robot market is growing rapidly. The *Silicone Robot* moves its head, mouth, eyes, and blinks while also answering some questions. The upcoming launch of *Doll Sweet* promises to replace *Emma*, who has been on sale since April 2017 (Owsianik 2017). Sex robots have not yet arrived in Italy, but one can now find a place open in Turin where customers can have paid relationships with silicone dolls (Ricca 2018). They are not yet sex machines but in other cities, robot dating houses already exist (Devlin, Lake 2018).

Very often, sex robots are presented as a product aimed exclusively at lonely people, the elderly, and those with severe disabilities (Di Nucci 2017). In reality, sex robots can be a fun 'toy' for any person (of any sexual orientation and age) who is interested and able to have sex with a machine programmed and designed for it (Bendel 2016, p. 18)—or who likes to have sex with another person using sex toys. It is a product that, for now, can only be purchased from online sex shops. Sometimes, sex robots can also hold (extremely) simple conversations in an enticing voice. Regarding the importance of the voice, "one should not forget that verbal eroticism is very popular in chats and that telephone sex was once in great demand and still exists" (Bendel 2016, p. 18). In other versions, the robot can also move and react to physical contact with a human being (Couric 2018).

Then, there are the sex robots that are mass-produced with the same appearance. Others are modular, according to the tastes and preferences of the buyer who can select hair color, eyebrows, and the shape of the lips, eyes, body, and genitals. The sex robots do not reason and have no self-awareness, at least for the moment, but they retain some ability to react and process external signals. Because they take a human form and have some ability to move, they represent a more advanced version of the so-called inflatable dolls and silicone that have accompanied adolescent fantasies of past generations. "Sex robots," John Danaher explains, "are different from traditional sex toys and other artifacts for stimulation which tend or replicate certain parts of the specific body, and thus do not have a human form, or which despite having a human form, generally lack any degree of artificial intelligence" (2017c, pp. 72-73). But even the sexbots of today are sex toys. They cannot replace a partner and we cannot build a relationship with them. But we can have fun, achieve pleasure and, if we want, also have sex with them either alone or with our partner.

How many people could ever be interested in having sex with a robot? In 2013, most people surveyed by the *Huffington Post*/YouGov did not seem particularly enticed by the idea of taking a robot to bed: only 9% of the sample imagined it possible, while 81% found it unacceptable; 10% were not able to give an answer at all (*Huffington Post* 2013). Four years later, another research study gave us completely different results. From a sample of 229 heterosexual males, 40.3% could imagine buying a sex robot within the next five years (Szczuka & Krämer 2017). One year earlier, research was even more favorable to sex robots as two thirds of males did not rule out the possibility of having sex with a robot (Scheutz, Arnold 2016). According to the latest survey conducted among Americans, 49% of respondents expect that in the next 50 years sex with a robot will become normal. However, only 1 out of 4 men (25%) and 1 out of 10 women (10%) reported that they would have sex with a robot (YouGov 2017).

Is the idea of a person having a sexual relationship with a robot (as intelligent as we want) really bizarre? After all, isn't sex different? We have sex with the person we love. Of course, we can also have sexual relationships with people we have known for a long time or even with people that we have just met and do not love. Likewise, we can just as easily have sex with people we do not even want to know. But how can we really think of having sex with a robot? It is true that we are talking about *sex* robots and their name allows us to imagine great things. Yet they are only machines, and sex evokes bodies embracing each other, mouths that kiss, hearts that beat, moods that are shared. However, things are more complicated than they may appear at first glance.

Let us try to better understand what sex is. What do we mean when we use this word or talk about a sexual relationship? The question may seem superfluous because many of us practice sex regularly and know when to do it and when to do other activities. Even so, this does not mean that we need to think about the concept or that we would be able to give a precise definition. The exact same thing happens with many other things that we do every day. How many of us would be able to explain what morality is and how (with reason or feelings?) we are able to distinguish correct behaviors from dishonest ones?

By sex, what we mean is that set of activities (everyone can think of what they prefer) that have a tendency to satisfy our desire for sexual pleasure (Primoratz 1999, p. 45; Morgan S. 2003; Morgan W. 2016; Webber 2009). As David Hume mused, "that under the term pleasure, we comprehend sensations, which are very different from each other, and which have only such a distant resemblance, as is requisite to make them be expressed by the same abstract term" (Hume 1960, p. 472). However, this does not mean that we are unable to distinguish one pleasure from the other. "A good composition of music and a bottle of good wine equally produce pleasure; and what is more, their goodness is determined merely by the pleasure. But shall we say upon that account, that the wine is harmonious, or the music of a good flavor?" (Hume 1960, p. 472).

Likewise, sexual pleasure is also easily distinguishable. It is the pleasure "that we feel in the sexual parts of the body: that is, both in the genitals and in the other parts of the body that distinguish the sexes" (Primoratz 1999, p. 45). This conception starts "from the fact that humans experience intense physical sensations through sexual activity that are extremely pleasant and embraces our commonsense opinion that it is precisely these sensations that are the object of sexual desire" (Morgan S. 2003a, p.1). This reconstruction of sexual pleasure may seem reductive, as we can even feel sexual pleasure in parts of the body other than the genitals. But the pleasure that is felt when we kiss or caress another body can be traced back to the excitement we feel in the genital area. "It is true that a kiss or a caress made only for the pleasure it gives the agent must not be accompanied from an excitement of the genitals ... But we must add that if it is sexual pleasure—and we experience some form of excitement—then this excitement will occur in the sexual parts of the body" (Primoratz 1999, p. 45).

But be careful: one can have sex without reaching pleasure. We might expect to spend a beautiful evening with someone, but instead things go differently. Perhaps we lacked a proper understanding of our partner's needs or there was not enough intimacy. We try and try again, but then we realize that the best thing is to stop. It is better to cook something, smoke a cigarette, or go to sleep. Next time will be better but this time, we receive no satisfaction from sex. We can also have sex not because we have a sexual desire, but because we want to have a child or we think we have duties towards God ("I don't do it for my pleasure, but to give God a son," recites the adage embroidered on the nightgown of Salina, the noble wife of Prince Leopard). There are people who have sex only for personal gain, because they hope to achieve economic or important advantages and unexpected career or relationship opportunities. Clearly, there are many reasons why we have sex. When we do it, we don't always seek pleasure; indeed, sometimes pleasure can be the last thing on our minds (Meston, Buss 2007).

However, we are accustomed to associating sex with pleasure and considering as sex any activity or practice that gives us sexual pleasure. In these terms, it is more than justifiable to talk about sex robots. Why should it not be? It is true that in the relationship with a robot, autoeroticism is exclusively at stake (as if the robot were a sex toy). But with autoeroticism, do we not reach pleasure? Woody Allen was right: masturbation is sex with someone we love (*Annie and I*, 1977). Throughout history, various arguments have been advanced to support the alleged relational nature of sex and the "perversion" of masturbation or, in any case, the impossibility of considering masturbation authentic sex. Yet the allegedly unnatural character of autoeroticism is highly questionable as how many people still perceive their bodies to be "forbidden territory" (Scruton 1986, p. 319)?

In view of its diffusion, masturbation seems a natural practice. One can also have very different opinions about what is natural. Therefore, describing something as unnatural does not conclusively demonstrate that it is truly a deviation or anomaly. Moreover, these considerations would have no moral value as something can be natural but bad, or not natural but good. Masturbation does not even seem to violate an alleged end of sexuality, as sex is often an intrinsically pleasant activity. So, it is incorrect to describe masturbation solely as a means of achieving other purposes (Goldman 1977). This is confirmed by science, which has shown the falsity of a finalistic conception of nature. In other words, it is not true that the sexual organs have their own natural function connected to a specific purpose (Casetta 2013). Add to this the fact that a relationship with another person is by no means an essential element of sex. Many relationships that we would not hesitate to define as sexual do not require it. I am thinking, for example, of the "the type of sexual experience sought by people who hang out in bars or on street corners in wait for a pick-up, go to sex-shows, resort to prostitutes, or peruse pornography. And it is equally out of touch with much, or most, sexual experience of many pre-literate cultures, or of more developed but strongly patriarchal cultures where marriage is typically polygamous." (Primoratz, 1999, p. 27)

We also know how easy it is to make love with a person while, at the same time, fantasizing about another person who is not present ("crazy idea to make love with him pretending to be still with you! Insane, insane, insane idea of having you here; while I close my eyes and I'm yours...", sang Patty Pravo). In this case, sex is present even if the relationship with the other person is missing (physically it is present but, as we said, it is absent in our thoughts). Further, it is not true that the other never exists in autoeroticism. The other person may be physically absent, but present in our thoughts and fantasies. And it can also happen that the person imagined in our erotic fantasies is much more present than the person with whom we make love. The other can also be physically present in autoeroticism. For example, partners may enjoy watching the other masturbate. The other may be present on the other side of the phone or video, or in a virtual way as an avatar. In conclusion, there is an evident continuity between sex and masturbation. Any attempt to distinguish one from the other is doomed to fail.

Using this, we can respond to those who claim that it is not possible to have sexual relations with a robot because sex is relational in nature—whether as an expression of intimacy and love, aimed at reproduction, or enabling the communication of feelings and/or thoughts. We can have paid sex with unknown people, either remotely over the phone or by having sex in virtual reality. Likewise, we can very well have sex with robots. It is true that we cannot have the same relationships with sex robots that we have with other living beings. It is difficult to be able to love a robot and the robot cannot reciprocate our feelings. But if autoeroticism is sex and sex robots really exist, then the time has come to take them seriously.

Is it moral to have sex with a robot?

Having ascertained that we can truly have sexual relations with robots, we can now ask ourselves whether having sex with a robot is morally acceptable. Even before the act, the very thought that one would seek sexual pleasure in a car (with a car) may seem repulsive to everyone. But if we want to reason about the morality of robots programmed for sex, we must first ask ourselves if there are sexual preferences that are more moral than others. Are some tastes immoral? Can taste say something negative about someone's character? For obvious reasons, sex practiced on non-consenting persons is not acceptable. For this reason, violence is something we condemn, no matter the victim. But things are different if sex concerns and is practiced between consenting people. In these cases, some preferences may appear eccentric or repellent but the fact that they surprise us is not enough to prove that they are immoral. To those who take a critical attitude towards certain preferences, one could answer that everyone is the best judge of their own pleasures.

The fact that others cannot grasp the pleasantness or pleasure of certain practices is only further confirmation of the fact that we are different people who build our identities and our lives according to patterns that do not always converge. This is the answer that could be given to those who reject same-sex intercourse as unacceptable, for example, or who deem it scandalous that people may find it exciting to have sex together. Some think sexuality is a practice that should only concern two people of a different sex with the aim of reproduction. Perhaps convinced that reaching orgasm with a sex toy constitutes a very inappropriate use of one's body and genital organs, they would say that toys should not be used to produce pleasure—only to have children. However, we live in secularized societies that are less and less marked by religious morality. The fact that one may want to enter a sex shop and buy a sex toy for personal use is no longer as scandalous as it once was.

Indeed, these toys are often a great idea for a birthday present or an engagement/wedding anniversary.

After all, what can be morally questionable about seeking pleasure using objects that resemble the body or parts of another person's body? No one suffers from the act and we can pass pleasant moments that help us overcome the stresses of daily life or prepare us to better face the difficulties and responsibilities of the next day. These considerations could be enough to show that there is nothing morally questionable about wanting to buy and have a relationship with a sex robot. At first glance, having sex with a robot may seem something different than reaching an orgasm with a vibrator or plasticized vagina. It is, however, still masturbation. If touching is not a sin and it is not condemnable to do it using toys or pleasure tools, then even sex robots cannot be something to be ashamed of. This is because "a robotic sex doll with only limited mechatronics and low-level AI is just a very elaborate act of masturbation, the ethics of their use will depend on the ethics of self-gratification" (Sullins 2012, p. 401).

We have already seen that autoeroticism is not a deviant or perverse form of sexuality. It is not true, in fact, that sex can be practiced only in relation either to another person or solely for the purposes of reproduction, loving, or communicating feelings. For a long time, masturbation was considered something unnatural and therefore immoral – sometimes even more seriously so than sexual violence and incest (Primoratz 1999, p. 43). Violence, adultery and incest are violations of human rules, said Thomas Aquinas, but they are still acts according to nature. However, masturbation is similar to homosexuality and bestiality in that it violates the natural order and, on these terms, it offends God who is its author (Primoratz 1999, p. 43). Others say that autoeroticism is a form of negating relationships with others. It is an unjustified interruption of the interpersonal relationship and/or an inability to manifest love and feelings for other people (Scruton 1986, p. 38). None of this is now defensible as there is no natural order to be respected, nature was not created by a superior being, and morality is a human product.

It is not true, then, that masturbation is an offensive gesture towards others. In autoeroticism, it increases one's pleasure and does no harm to anyone (Levy 2008). You can have fulfilling and respectful interpersonal relationships and, at the same time, allow yourself the pleasure of masturbation when you feel like it. One thing (masturbation) evidently does not exclude the other (relationships). Some believe only actions count in ethics and that an action is all the more just (morally more appreciable) the more capable it is of promoting the well-being and happiness of the people involved. From this perspective, we cannot criticize a person for having sex with a robot. We might not do it in their place, but this person does nothing wrong. However, those who defend an ethics of so-called virtue would not stop at this conclusion. Instead, they would likely ask us to further clarify and study this issue. For supporters of virtue ethics, a person's morality is not measured solely in terms of their behavior, but also in terms of their reasons. As Alessio Vaccari explains, it is measured on "personality traits that do not end with the action but survive it and become stable principles of conduct" (Vaccari 2012, p. 23). That is, behavior has no merit because the object of our evaluation is character (Hume 1960, Book 2 and 3). It is true, then, that those who have sex with a robot do not harm anyone. Unlike humans, sexbots cannot suffer violence, they cannot share our pleasure, and thus they cannot even suffer. (They could be damaged during sexual intercourse but one can buy spare parts just as easily to have them repaired, or replace them with other robots. In any case, damage would be an act of vandalism – not violence.)

For supporters of virtue ethics, however, this is not enough. Instead, we should ask ourselves what kind of people they are and whether their motivations (that is, what pushes them to have sex with a robot) are truly appropriate. If we are able to look at things from this perspective, according to some supporters, we would inevitably start having doubts about the morality of sex robots (Sparrow 2017). Is it not logical to think that behind the choice to buy or have sex with sex robots, there is an underlying difficulty in relating to other people? Is it not evident that behind the need to buy and own a robot, there is the desire to have a completely submissive (Gutiu 2016) and controllable partner (Sparrow 2017) – one who is always available and does not demand treatment and attention?

It does not take much imagination to get an idea of the type of person who might find it exciting to buy and have sex with a robot. According to Richardson (2015; 2016), the fear of abandonment turns into hostility towards women over time, pushing men towards sex robots. If men had the ability to have mature relationships with women, they would not need robots. But relationships pose a threat to them. If anyone still has doubts about whether these are people with a serious deficit in moral sensitivity who are deeply depraved, says Sinziana Gutius, it would be enough to see the SBS documentary *My Sex Robot* (2010). These are the same men who admit their problems without shame: "the fembot enthusiasts said they were especially drawn to the idea that sex robots would not challenge them in the way that a human relationship would, and preferred sex robots as less emotionally risky entity. They expressed discomfort at the idea of interacting with women, and shared stories where their feelings were hurt because women rejected them. One of the men featured acknowledged he had trouble engaging women romantically after a difficult breakup, and stated that being with a sex robot is 'better than seeing a shrink.' Another man featured in the documentary said that sex robots meant 'being with somebody without all the strings that come attached to being with another person'" (Gutiu 2016, p. 208).

But how do we derive precise and reliable conclusions about a person solely from what we know about the sex toys they use or buy? It is complicated enough to understand a person by his or her sexual preferences or habits. Would someone have the courage to argue that there is something wrong with, for example, a person who enjoys anal sex with pleasure – that "normal" people should only practice vaginal intercourse? (This, by the way, would condemn any gay couple or relationship as immoral). And if we want to refer to other sexual activities, would someone say that a person who does not like oral sex is better than someone who practices it? Tristan Taormino is right when he says the idea that anal sex is vicious, abnormal, or perverse "is based on the assumption that only one form of sexual expression – specifically the heterosexual interaction between penis and vagina – is natural, normal and conventional. Any other activity, including manual stimulation and oral sex and sex toys, is considered abnormal" and a sign of a non-exemplary character (2017, p. 22). This position on sex was once widely accepted at both the level of public morality and legislation that, for example, enforced severe penalties for sodomy. But today, would it still be possible to support such laws?

The purchase of a sex robot is not necessarily the sign of an inadequate moral character. Even a virtuous person with morally appreciable character traits, what we would commonly call a "beautiful" person, may buy one. It is true that there may be cases where the only motivation for having sex with a robot is the difficulty or fear of having a relationship with other people, or the existence of a hostile attitude (whether conscious or unconscious) towards women. A person can treat the robot as a substitute for a human partner only because they are already accustomed to considering their fellow humans as a means. "A real human lover," writes Hauskeller, "can be replaced by a robot without any loss if, and only if, other people can never be more than a means to us, if they already are, for all intents and purposes, merely sexbots in disguise" (2014, p. 14). However, we have no reason to think sexbots will be attractive only to those people who are no longer able to have equal relations with the other sex and who would like to have a woman who is always passive and submissive. Sex robots may also be of interest to individuals who do not feel the need to find a partner to fulfill their every desire, or who do not live in terror at the possibility of a woman refusing their advances.

In fact, the choice to buy and have sex with a robot can come from several reasons. For example, one might do so from the desire to have an exciting object for moments of pleasure that is more interactive than other toys currently on the market intended for auto-eroticism. There could also be curiosity behind the purchase of a sex robot – the desire to experiment with new opportunities for pleasure with a partner, or to liven up a relationship that has slowly slipped into habit and boredom. This attitude can also be the sign of an appreciable character (after all, some people may be flattered or appreciate the fact that their partner has a desire to do new things with them). Others, however, may be disappointed by a previous relationship and not feel ready to start a new love story, still being in love with their ex. Some of them would probably like to have a more active partner and sex life but unfortunately, for various reasons, they are unable to find it. Here, we must not only think of people with disabilities as there are many circumstances that can make it difficult for a person to have significant emotional relationships. One can imagine, then, that sex robots could also target someone with a partner who is regularly abroad or who, for some other reason, cannot have sex at that time (who suffers from an illness, for example). Finally, robots could help people who are shy or depressed, living with sexual trauma, or putting a recent breakup behind them overcome a difficult moment in their life. With help from a robot, people with disabilities or malformations could return to having a satisfying sex life without the need to resort to prostitution; less experienced ones could improve their amatory skills and thus prepare themselves for a more positive overall sex life (Di Nucci 2017; Appel 2010, pp. 152-154; Casalini 2013; Seibt et al. 2014). There are still no sex robots on the market that can teach us to become better lovers, but someone will surely think of designing them. Through such sexbots, people could more about sexuality, where and what the genital organs are and look like, how they work, and how pleasure is achieved. They can also explore positions they have never imagined or experienced.

Whether the purchase of a sex robot signals a morally appreciable character (or person) depends first on the reasons motivating this choice, then on the consequences it produces for other people directly involved. The reasons may be reprehensible, but they can also be laudable. Behind this choice may be a sincere concern for the pleasure of other people, for the stability of one's relationship, and also for one's well-being. After all, having a satisfying sex life not only contributes to our happiness, but also has positive effects on our health as there is a correlation between an intense sex life and weight loss, lower stress levels, better heart health and blood pressure, reduced risks in contracting prostate cancer, and the ability to fall and stay asleep more easily (Whipple 2008; Brody 2010; this research is cited by McArthur 2017, pp. 43-44). However, sex with a partner has greater benefits than auto-eroticism or masturbation. "We do not fully understand the reasons for this," Neil McArthur writes, "and so we cannot say whether robot sex will achieve the same benefits as partnered sex. However, I think it is plausible to say that sexbots will deliver at least some of these benefits. For instance, robots will require the same level of physical exertion as sex with a human partner. Indeed, they could be programmed to require more. And the touch and feel of another person seems to activate a certain sort of physical reaction in us that (at least as the technology develops) may be achieved with a robot" (2017, p. 34). It is understandable, then, that there are important concerns about sexbots and also strong negative feelings about the idea that someone may want to embrace, kiss and have sexual relations with them: we are dealing with new technologies that open up completely new scenarios. The impulse to reject new things is a normal one, and there may also be reasons to doubt their acceptability. But it would be wrong to dismiss the matter by stating that only people with serious deficits in moral sensitivity have the desire to buy them. If we do not like sex robots and want to show why it is wrong to use them, we should highlight problems other than those we have considered so far.

Can rape robots be a game?

The frame of our reflections does not change even if we take into consideration sex robots that are programmed to reject a sexual relationship or to express disgust or little appreciation for any sexual advances. In the past, True Companion advertised a version of Roxxxy called, not by chance, *Frigid Farrah* to allude to her lack of interest in sex. Recently, Synthea Amatus introduced *Samantha*, the first sex robot who says "no" if she does not feel like it (Miley 2018). Imagine, then, someone continuing to try to have sex with a robot after it says "no." Perhaps the robot could physically withdraw, shout "stop, you are raping me. Don't do it again" and fight to end the abuse (Sparrow 2017, p. 468). What would we say about such a person?

Robots do not experience emotions and feelings. They are not even self-aware; we are still a long way off from being able to interact with intelligent and self-conscious robots. But according to Robert Sparrow, the fact that a person would want to have sex with an "unwilling" robots (that is, one programmed to refuse any sexual approach) tells us we are encountering a truly hideous person. This person enjoys imagining violence against other people. They buy and have sex with robots programmed to reject any sexual relationship only because they take pleasure in living experiences of violence and harassment. The problem is not that the likelihood of having an irresistible desire to rape "people in flesh and blood" increases by dint of "raping" a robot, or that the same actions will then be committed on women and children. As we will see, this is also an important question raised by those concerned with producing or selling sex robots and that is the motivation underlying the request for an international ban on these kinds of products.

The point that is sustained here is another: the fact that it would not be possible to get excited by reciting or imagining violence and, at the same time, be a virtuous person. This inclination would reveal that one is sexist, intemperate, and cruel (Sparrow 2017, p. 473). Violence would be practiced only on robots, but it would still be violence. The pleasure that accompanies it would be an indisputable sign that we are dealing with a depraved person. In short, "raping" a robot would be wrong not because it spoils the character of a person ("when we participate in simulations of excessive, indulgent and illicit actions, we cultivate the wrong type of character…you hurt yourself because you erode your virtue and in this way you move away from the goal of perfection," McCormick wrote in 2001). It is because the action itself is an expression of a vicious character (Patridge 2011).

But is it true that only a cruel, violent, and sexist person might want to buy a sex robot programmed to reject any sexual advances? At first glance, Sparrow's hypothesis seems convincing. Yet he does not seem to take into account that while it is one thing to draw conclusions about character from actions, it is another to reconstruct a person's character starting from their fantasies. From what a person does, I can get an idea of who a person is. But are we sure that their fantasies also tell me something important about their character? Maybe some people would buy a robot for non-consensual sex tomorrow because they sometimes have a desire to enact violence. Doubtful of causing suffering to another person, perhaps they would take no pleasure in having a sexual relationship with a robot that refuses their approaches. But robots cannot suffer and therefore have no moral scruples. In this case, would we say that we have degenerate or vicious people before us?

Moreover, this would not be the first time we have practiced violence against another person within a world of fiction that is clearly distinct from the real world. We have always done so in theater and in novels, as well as in private life. Is this not the attitude between consenting adults who enjoy having sexual relations in which they act out scenes of sexual violence? One of the most popular and most accessible objects in sex shops are handcuffs. Online sites advertise all kinds of handcuffs from feathers and silk to metal. They can even be purchased at a paltry price on Amazon. Lovers of bondage cannot do without collars and ball gags (a ball gag is usually a rubber or silicone sphere with a strap passing through its diameter, according to Wikipedia). For those who prefer more extreme bondage, there are constricting ropes. For him, there are chastity rings to squeeze the testicles. For her, there are those to pinch the nipples and clitoris. In short, a variety of objects serve to mimic a series of erotic practices involving pain and pleasure through a relationship of domination and submission. There is even an abbreviation to define them (BDSM, an acronym that literally stands for Bondage, Discipline, Submission, Sadomasochism; those who have read the Marquis Sade's novels or *50 Shades of Grey* know what we are talking about).

Is there really something morally repugnant about playing these games? Are we sure that only the most morally corrupt people could achieve pleasure by having sex with their partner after tying them up and/or whipping them? In the event their partner has not consented, we would have no doubt; it would be an inadmissible violation of their rights and an insult to their dignity. But here, we are talking about people who have agreed to be part of the game and who could stop the other at any time by saying a safe word (in bondage, in fact, a "safe" word is established from the onset that can be said be a partner who wants to stop the game). There is no compulsion as everything happens consensually and with complete freedom.

Violence is also enjoyed in another ways, such as through the use of pornographic material, novels, films, and pictorial representations. Not all pornography represents scenes of violence, domination and submission, but the representation of violence is certainly a recurring theme. For this reason, pornography has been bitterly contested. Some have argued that it should not be allowed because the representation of female performers as an object harms their dignity. Of course, if this were true, people who buy or consume pornographic material or who regularly watch pornographic videos or other material (from photographs to drawings as now you can find everything) on freely accessible online sites, would show low sensitivity for well-being and women's rights. They would think only of their pleasure and not worry in the least about the consequences of the production of pornographic material. But as we will see in the next chapter, it is debatable whether pornography is harmful to women – whether it actually increases the likelihood that they will then be raped and subjected to the degrading practices shown in red light films.

The fact that a person feels pleasure in consuming pornographic material that represents scenes of violence is neither necessarily the sign of a violent character nor indicative that they are a person without any scruples. "The object towards which the excitement of the users," explains Carola Barbero, "are the characters who take part in the pornographic representation which, as such, is a genre of fiction. It is therefore a reaction directed towards fictitious characters who, from an ontological point of view, are objects of the same type as *Madame Bovary* or *Mickey Mouse*. Usually, the user is well aware of the fictitious status of the objects that provoke such an emotional response in them" (2013, p. 144). That is, the context in which the pleasure of violence is sought is important.

The enjoyment of representations of violence here is not the same thing as the enjoyment of real violence because the objects of violence, whatever they represent, are not real. Moreover, fiction allows things that reality does not. Fiction can be a sadomasochistic relationship or a scene from a pornographic film – it makes no difference as it is not true violence, but pure fantasy. "[It is] one thing to say to a woman, in a pornographic film, 'You are my slave,' and a completely different thing to tell them that on the street. In fact, if in the first case the profusion of the statement takes place within a context of fiction in which the subject is recited, in the second case, the statement is used in a real context. And it seems that only in this second case is the submission and violence of women's rights real. Evaluating a profusion of the context of fiction as if it belonged to the real context is not a correct move, and this is trivial because fiction admits a profusion that reality forbids" (Barbero 2013, p. 146).

From what we know about people who buy and play violent video games, we have no reason to suppose a necessary desire to do violence behind the play of violence. Some people will undoubtedly have this motivation and we do not want to deny it; they choose to play certain video games and are not interested in anything else because they take pleasure in imagining causing suffering and really killing. But others may be fond of violent video games for different reasons. Perhaps they are attracted to graphics or the challenges that the games entail. Morgan

Luck details how when chess is played, taking the opponent's pawn represents a defeat for the enemy army. But the chess player does not take pleasure from the representation of the killing. Their pleasure derives from something else, such as from winning the game. "The same might be said for virtual murder within computer games. A player may enjoy a computer game because, for example, it satisfies her competitive nature, not because it allows her to commit acts of virtual murder *per se*" (Luck 2009, p. 34). What applies to murder, then, could also apply to virtual pedophilia (to those games in which the characters represented are not real, but virtual): "To illustrate this point," Luck explains, "imagine you are playing a computer game, the object of which is to steal the Crown Jewels from the Tower of London. One way to achieve this goal is to seduce and sleep with a Beefeater's daughter, who just so happens to be 15. A player who commits this act of virtual paedophilia may do so, not because he enjoys the notion of having sex with a child, but because he wishes to complete the game" (Luck 2009, p. 34).

It might also be admitted that some video games allow the player to perform actions that would be considered immoral in reality – ones that do not increase their chances of winning or gaining more points or fuel reserves (such as running over pedestrians who stroll down the street). When playing on a computer, there is probably some pleasure in killing and doing other terrible things such as torturing, raping, or inflicting suffering, as "game makers have made some of these games more and more graphic in their portrayals of torture, assault, murder, and other acts of violence. Whereas shooting an opponent from a distance would have once resulted only in the collapse of his or her body, now the shot is accompanied by screams of pain, realistic writhing, blood, specific damage to a part of the body, flying body parts, and death" (McCormick 2001, p. 277). But do we want to conclude that these people have an evil character? Are they indeed people who cannot adequately sympathize with the victims of their violence and display a serious deficit in moral sensitivity as a result? Given the spread of video games that at least involve killing innocent people without suffering, this would be really worrying (but video games can also include much more monstrous actions). But a different interpretation is also possible: in line with what has already been said, it could be argued those people are so immersed in virtual reality that they do not think about what their actions mean for the real world. For them, killing in front of a screen is not really killing a person.

It warrants asking ourselves what would be different about purchasing and then having a sexual relationship with a sex robot programmed to be non-consensual. If using a character or avatar to kill and rape in a video game or virtual reality is different from doing the same things in reality, why should raping a robot be like raping a real person? According to Danaher, what makes committing violence on a sex robot programmed to refuse sexual intercourse different from enacting rape in video game would be the fact that the relationship with the sex robot does not provide any distance: "In the case of the video game, there is some physical and mental distance between the game player and the virtual representation of what they are doing. This distance will always make it more difficult to draw inferences about the player's actual moral character. In the case of purely robotic acts of rape and child sexual abuse, this distance is massively reduced, and it is consequently easier to make those inferences. We can be much more confident (than we can be when we consider the same violence perpetrated in a virtual context) that the general act-types involve harm to moral character, or include wrongs done to the public at large, than we can be in the virtual case" (2017c, p. 86).

Considering the distance between our world and virtual reality, Danaher says the video game allows people to act as if they were other people (the character they are playing with). In the case of the interaction with a robot, however, this distance would be missing. Therefore, the agent could not be someone else; it would be themselves, with the consequence that their actions would be a direct expression of their character. In having sex with a robot, the interaction would show who they really are: "The purely robotic act does not involve action through an image or other medium. It involves direct physical contact with an artificial being. That means the personal distance between the act and one's true intentions or desires is lessened in this case. It is not as if you are someone else when performing these acts; you are yourself. Hence, we can be more certain of the inferences we draw about the moral character of the person who engages in those acts" (Danaher 2017c, p. 87).

However, Danaher's considerations can be answered in that the artificiality of the robot still allows the people involved to create with the robot the same distance that the player has in virtual reality. Yet according to Danaher, such an argument cannot be truly convincing because we have research clearly attesting that it is more difficult for people to perform immoral acts with physical representations of real things than to do those same things abstractly or in a virtual reality. The fact that people might have trouble hitting the head of a toy child on the table or hitting another person on the knee with a toy hammer – even though we know the child is not real and the hammer cannot hurt – it would make a difference for Danaher, given that we are "programmed" to react in this adverse way. That is, we are programmed to have an "immediate" or "intuitive" moral response of this type to the thought of causing harm to a person or something that resembles a person (Danaher 2017c, pp. 87-88). So, if someone did not feel this same feeling of repugnance or resistance, they must suffer from some moral deficit: "My belief is that this study has something of interest to say about the cases of robotic rape and child sexual abuse. More precisely, I believe it suggests that those who perform and enjoy performing purely robotic acts of rape and child sexual abuse must either (a) have an inherently defective system of intuitive moral judgments or (b) have worked to repress or overcome the intuitive resistance to such acts. In either case, there is reason to think that actions of this type are harmful to their moral character and are hence apt for criminalisation" (Danaher 2017c, p. 88).

However, it is not at all clear that the results of this study or similar research are sufficient to show the moral inadequacy (or deficit, as Danaher calls it) of people who choose to play with unwilling robots. First, our most immediate reactions are not necessarily those that can claim the greatest moral authority. Our most immediate or intuitive reactions are authoritative and therefore count from a moral point of view only if they can be accepted and approved reflexively – that is, if they pass the critical examination of "reason" with flying colors (Lecaldano 2010). This means that sometimes the resulting reactions are correct, but other times they can only be the expression of unjustified prejudices. The fact that we respond to certain situations in the way Danaher suggests does not mean that this is the right way to deal with things.

Our reactions can be influenced and conditioned by aspects that should remain outside our observation and evaluation of situations. For example, we might have a moral scruple to perform those actions that do nothing to a doll but would be deadly for a child. These reactions, however, are completely irrational and cannot be morally justified. It is difficult for us to hit the head of a doll on a table or inflict other terrible tortures on it only because we have the impression that we are dealing not with an inanimate object, but with a living being capable of experiencing pain. Perhaps, then, the doll reminds us of one we played with as children and grew fond of like a sister. She slept with us, we took her to school and even on vacation, and we were inseparable. We also gave her a name and it seemed like more than just a doll, which is also why it can be difficult for us to think of doing something wrong to it. But a doll is generally made of plastic and fabric or sometimes wood and porcelain. It is an object; it cannot suffer, and we can therefore treat it as we want. From the fact that some people treat it for what it is and not as a person, we cannot draw inferences about their character and their sensitivity. Perhaps they are simply able to keep fantasies (that is, fiction) separate from reality better than most people can.

Furthermore, Danaher's argument draws conclusions that are difficult to accept as it seems we should disapprove or otherwise morally criticize not only people who choose to have relationships with unwilling robots, but also those who play certain roles or characters in theater and cinema – and who, for this reason, taint themselves with figurative crime scene and commit violence. Even these people would appear morally insensitive to Danaher, as they are able to mimic and therefore perform certain immoral actions. They perform these actions not through an avatar or a medium, but directly. This is to say nothing of the people who, as we have mentioned, love to have sex with their partner by mimicking scenes of violence and abuse. Even these people would, according to him, show at least suspicious dispositions as they would still allow themselves to commit actions we disapprove of.

Additionally, the criterion adopted by Danaher to distinguish between violence perpetrated at a distance and violence practiced at close range is not very clear. Some types of virtual experiences can be particularly real. If we have violent interactions in such cases, are we still carrying out violence at a distance? (These are experiences that, for Danaher, could allow people to act assuming the point of view of the first person rather than the third person). For example, would having sexual intercourse through a teledildo and imagining raping the other person be remote violence or close violence? The act of sex is practiced remotely because the other person is on the other side of the screen. But by wearing equipment, one can provoke and transmit real sexual stimulation to the other person (teledildonics) via the Internet. Moreover, the experience of proximity and distance can be very subjective. An object can be equidistant but perceived differently by people. It is not only the position occupied by the object and the distance between us and the object that counts, but also the intimacy and knowledge we have of the object – and perhaps also the type of value that we attribute it. For this reason, it may be very inappropriate to assume distance and closeness as a criterion for a moral evaluation that is purportedly objective, at least in terms of the character of people who want to have relationships with sex robots.

We started with the thesis put forward by those who argue that only morally corrupt people might want to have sex with a non-consenting robot. This criticism is not advanced towards sex robots in general, but only towards a particular version of the sex robots currently on the market. Even assuming that the argument was acceptable, it would still not be able to show that the purchase of sex robots is always an indication of a character that takes pleasure in violence (those who purchase sex robots who display pleasure in being given sexual attention, for instance, could not be criticized for having fun sexually abusing them). However, we have shown that there is no necessary connection between acting out a scene of sexual violence and a corrupt character. The possibility of staging sexual abuse or violence on a robot obviously does not arise with the recent production and marketing of sex robots. We can also practice this game with human partners. For example, we can ask our partner to play the part of someone who do not want to have sex and, together, we can simulate a relationship practiced against the will of one of the two parties. Furthermore, it is not necessary for the same person to always play the part of the victim: the parts can be reversed and those who used to do violence will once again suffer it.

We can play these games not just in the real world (in a bedroom or anywhere else we can imagine) with other consenting adults, but also alone. Through an avatar, we can enter virtual reality and play as an American military general or a pirate to commit violence against other people. The more violence committed against people, the more likely we are to win. To those who enjoy these things, sex robots seem to offer a new possibility: we may not even be able to imagine what is pleasant about playing to rape another person or letting another person play violence against us. These things may seem not just in bad taste, but even repulsive (Sparrow 2017, p. 473). They do not seem, however, the obvious sign of a moral deficit. Of course, we could legitimately ask ourselves what the consequences of these amusements could be on a person's character over the long term. We might wonder if there is no risk of them gradually beginning to have the desire to commit violence against real people. But we will address this in the next chapter, when we ask ourselves whether relationships with sexbots fuel violence and abuse against women.

Sex robots and self-deception

There is still something that people who enjoy having fun with a robot in bed do not like. We might agree that they do not necessarily have a bad character and that they may even get us used to the idea of these kinds of relationships with machines. But the point is that these people consciously choose to deceive themselves (Sparrow 2002). They treat these machines that are human in appearance only as a real partner – a person with whom they can have emotional relationships as well as sex. The companies that represent sex robots have no intention of deceiving us. The sex robot is not presented as a human being, but as a technological object and each of us is free to choose whether to buy it and then play with it. If there is deceit, it is a kind of self-deception on the part of the person directly involved who, even though they know they have a robot in front of them, still treats it as a partner.

This has already occurred with robotic animals. The uses of these robots in medicine and assistance for the care and companionship of sick, disabled, and elderly people help to construct their affective relationships with these robots, and thus the robots' mechanical and inanimate nature are forgotten. The same thing could happen with the sex robots on the market today. Particularly if a person is alone, they could become attached to the robot they purchased and begin to perceive it as a person capable of reciprocating their feelings. The fact that this happens is explained by our tendency to anthropomorphize machines and believe that they are capable of doing many more things than they really do. Even if we know very well that the object is inanimate, we are able to suspend belief and treat it as a human being. For instance, how many people do we know who talk to their car and speak consolingly to it after someone inadvertently scratches or damages it? We even encourage our computer when it fails to execute our commands and, after trying unsuccessfully, get to the point where we threaten it (Turkle 1984; Coeckelbergh 2011, p. 199). Our relationship with computers may go even beyond this, because we can consider it our friend and accomplice or give it a gender or a personality (Coeckelbergh 2011, p. 200).

If this is true for computers, it is all the more applicable to robots operating in the personal sphere. "Robots cross this inanimate-animate border easily – they often appear animate – and we interact with some robots as if they were human. If computers already give us the experience of animation, we can expect robots to do a much better job. Many (autonomous) robots appear animate. Whether or not their appearance is human-like, they can be equipped with more possibilities of expression in order to enhance the experience of animacy" (Coeckelbergh 2011, p. 199). For this reason, people who buy and have sexual relations with robots would fall victim to a deplorable sentimentality (Sparrow 2002, p. 306). They would build important relationships with them and feel feelings towards them that can only be experienced with sentient and conscious beings. In other words, people would be convinced that the robots are truly different from what they really are (not robots, but real humans) only because the former can abandon themselves to these feelings (Midgley 1979, p. 386).

In this way, they make experiences that, in the best of hypotheses, have no value. These experiences can be harmful as they are tied to a distorted perception of reality or false beliefs on an unconscious level. We can only feel love and affection for our robot if we believe that it returns our love – that it is flirting when we return home, that it reflects genuine happiness or has a bad mood and real sadness, etc. To maintain the same sort of Involvement with a robot that we can have with a human being, we must be wrong about its real abilities. Of course, Sparrow acknowledges, itis very likely that people will be well aware on a conscious level that their sexual partner is simply a robot. But if people claim to love their robot in the same way that they could love a real person or are keen to keep their friendship while knowing that it is simply an inanimate robot with programmed reactions, "to the extent that they are in fact reporting accurately on their feelings, these can only reflect the presence of mistaken beliefs at a subconscious level" (Sparrow 2002, p. 315).

We, too, can admit that continuous interaction with machines who look like real human beings can encourage people, on an unconscious level, to maintain the same feelings we have for other humans. It can also encourage a relationship of care and love that, at the moment, seems inappropriate for robots. We can recognize that some people will need, as Sparrow argues, to invest "sentimentally" in these machines because they do not have other people to love or because they are going through a difficult time. We do not want to deny that this can happen. It can happen with robotic animals, so there is no reason to think it cannot happen with sex robots. But this need not happen every time a person buys a robot. Even in this case, we should avoid letting go of easy generalizations that cannot be proven empirically. One can imagine that only a very limited number of individuals will need to confuse their fantasies and hopes with reality. Many other people will have no reason to invest sentimentally in robots. These people will already have fulfilling emotional relationships and will not feel the need to build other, artificial ones with robots who cannot reciprocate their affection. When needed, they will use robots as sex toys but then put them aside, forgetting about them until the next time they want to use a toy.

Even if we want to admit (without really conceding) that the use of sex robots as sexual partners necessarily gives rise to emotional relationships, why should we criticize people for their sentimentality? In interacting with a robot and in feeling some sense of love and tenderness towards it, would we perhaps fail to fulfill our duties towards ourselves? (In turn, would this mean those who make robots are morally objectionable because they encourage us to take what is only an illusion for reality and truth? That is, would they fuel our inclination towards self-deception?). We agree that robots cannot reciprocate our feelings and we can also agree that they are not "ideal" partners for an emotional relationship – that instead of deluding ourselves, we should look for people who really love us because the robot cannot give us the same love and the same care that a human being can give. But do we not have the right to sometimes pretend that reality is different from how it actually is? What would be wrong with imagining that it is more tolerable and pleasant? When reality becomes too harsh and painful, for example, are we not allowed to transfigure it – at least for the time necessary to rework our experience and find the ability to face it again for what it really is?

Even if we are not experiencing a difficult situation (we did not just have a hard breakup and a period of mourning or we are not going through a period of depression and loneliness), do we not have the right to give ourselves a break from reality and play with a sex robot by pretending it is not a machine but a human being? As Raffaele Rodogno states, "Sparrow believes that sentimentality is morally wrong when it violates the weak duty to ourselves to apprehend the world accurately. But what exactly would be wrong with failing to apprehend the world accurately, above all when by doing so no one else is being harmed and when there is a suspicion that a subject's acquiring or entertaining incorrect beliefs may actually be beneficial to her?" (2016, p. 263). A correct perception of reality has both an intrinsic and an instrumental value, allowing us to create the things that are important to us (Blackford 2011, pp. 45-48). However, it is not always wrong to indulge in false beliefs. Sometimes, for example, we are able to accomplish great things only because we overestimate our abilities and attribute to ourselves qualities that we do not really have.

This view of things can also be good for our health (Taylor & Brown 1988; Synder & Higgins 1988): "psychologists often claim that clinically depressed people have more accurate perceptions than others of their true abilities. Indeed, improbably optimistic beliefs about one's abilities, one's control of events, and one's life in general, appear to facilitate superior health, happiness, and performance [...] unrealistically positive self-evaluations, exaggerated perceptions of control, and unrealistic optimism are pervasive and enduring illusions that actually benefit mental health, the capacity for creative, productive work, and the ability to care for others" (Blackford 2011, p. 44). Yet we overestimate not only our own qualities, but also those of other people. When we are in love, for example, the other person seems to us to be the most wonderful in the world and we see things about them that other people cannot see.

For this reason, Smith writes, it is difficult to sympathize with those in love because their passion "appears to everybody, but the man who feels it, entirely disproportioned to the value of the object; and love, though it is pardoned in a certain age because we know it is natural, is always laughed at, because we cannot enter into it" (Smith 1995, p. 119). But what makes us lovable is the fact that we are capable of this idealization. If we saw things more rationally, we would have a more objective perception of reality. But we would probably lose the enthusiasm for falling in love and not have the same care or attention towards our partner.

We also deceive ourselves when we read a novel or see a film and sympathize with the lives of their protagonists, who appear to us to be real and authentic people. We feel those butterflies in our bellies when we have to separate ourselves from the toys of our childhood or from the things that have long belonged to our family: in this case, too, we attribute to things an exaggerated value that they do not have (Blackford 2011, p. 47). These reactions can also be criticized for their naive and easy sentimentalism. But the meaning of life is also in these things because without these feelings, life would certainly be less exciting and enjoyable. For this reason, "despite Sparrow's concerns [...] we should be lenient about familiar, relatively benign, kinds of self-indulgence in forming beliefs about reality" (Blackford 2011, p. 41).

Playing with a sex robot and treating it like a real person could help rid me of boredom and loneliness. It could also make me a little happier. It can allow me, then, to distract myself from concerns that trouble me and/or other people. It can help me distance myself for a while from people with whom I have to interact (because I need their assistance, for example), but with whom I do not always get along. At least until I take this relationship with the robot too seriously and as long as it does not become the only relationship I can afford or am able to build, there is no fear of negative consequences for my life or that of my loved ones. The fact that I play with this robot as if it were my partner does not mean that I will then no longer be able to deal with reality in the most correct way. It does not mean that I will stop working or begin to treat each machine as a person who deserves the same relevance and respect as a human being. In the social relationships that matter, I can still be a reliable person: "Smith might think of the plants in his garden with a degree of affection that they are quite incapable of reciprocating. He may subconsciously imagine that they are grateful, and he may be happier and healthier [...] his arguably sentimental attitude to plants does not entail that he consults them about family difficulties or hears them whispering advice about buying and selling on the stock market. In any context that calls for critical reflection, he will report that they are really just plants" (Blackford 2011, p. 47).

Our undeniable tendency to zoomorphize machines does not seem to compromise our ability to distinguish them from people. We may sometimes turn to our car as if it were a dear person in need of care and attention (who we have to refresh, clean, and feed). However, we would never put it on the same level as another human being. When it is necessary to do this, we are able to distinguish a thing from a person. Our attitude becomes worrying only if we lose this ability. At that point, we really risk confusing our fantasies with reality and not being able to take care of ourselves and others. But those who buy a sex robot and let themselves go into a first sexual and then sentimental relationship can also be aware that they are only interacting with a robot. They can interact with a robot animal or an android, taking pleasure and benefits from its company without having to perceive it as a real human. The presence of the robot can cheer up a person's day and provide an important opportunity for distraction or a stimulus against boredom and depression, but the sex robot need not be confused with a human being in order for this to happen (Coeckelbergh 2011, p. 199).

As in the case of fiction, we can let ourselves get carried away by narrated stories or become involved with the lives of the protagonists. We can do so without minimally believing that these characters really existed. In the same way, we can benefit from interacting with robots without having to seriously believe that the machines in front of us are alive. We can be emotionally involved with robots in a way that is similar to how we can be emotionally involved in a good novel or a good movie. In sympathizing with Anna Karenina, we imagine that certain unfortunate events have happened to her (Tolstoj 2013). In the same way, we can imagine (without really believing it) that the robot is happy to see us and that it experiences pleasure under my care and attention. In this relationship with robots, there is no sentimentality because there is no irrational or incorrect representation of reality. There is only a desire to play and take pleasure from this original interaction.

Even these considerations seem to further confirm the difficulty of considering sex robots as an object repugnant or contrary to morality. In the first part of this chapter, we saw that sex robots could enrich the possibilities of sexual pleasure for those who love to practice auto-eroticism and who do not disdain the use of sex toys with their partner. In the second part, we discussed the criticisms put forward by those who think that sex robots are purchased only by people with a moral deficiency, such as more or less conscious hostility towards women, various relationship difficulties, a desire to have a passive and submissive partner, a desire to discover the pleasure of violence and, finally, difficulty in taking reality seriously and a tendency towards self-deception. We have seen that none of these things are true. There is no evidence that behind the interest in sex robots, there must necessarily be moral provisions showing a lack of respect for other people or an inability to take adequate care of self and well-being. This is also confirmed by the latest considerations of the alleged sentimentalism that would outline the character of those who have relationships with sexbots. After all, what would be so questionable about having a sex robot next to you? And why should these robots even be dangerous for us? Contrary to popular belief, sex robots could prove to be an important resource – not only for sexual pleasure, but also for companionship. In this regard, we have not yet adequately investigated the possible benefits of sex robots and therefore it is certainly premature to express a definitive judgment about their usefulness. But it is possible they have benefits similar to those obtained from pet robots. An increasing amount of research shows that people who interact with a robot animal can benefit significantly from its "companionship"; awareness that they are not living beings does not negatively affect the experience of wellbeing and happiness (Banks & Banks 2002; 2005; Banks er al. 2008). Positive results have been observed in terms of reducing loneliness in older people, among other things; important benefits have also been recorded in those suffering from dementia, including an improvement in communication (Bennett et al. 2017, pp. 536-539; Robinson et al. 2013; Tamura et al., 2004; Kanamori et al., 2002).

As has been pointed out, the most interesting thing is that these benefits would not be different from what a patient can get from interacting with a real dog (Bennett et al. 2017; Bemelmans et al. 2012; Banks et al. 2008, p. 176). However, it will be a matter of better understanding if people interacting with a robot could have even more important benefits in a friendlier social environment. Furthermore, we do not know whether the benefits of robots last over time. Playing with one may seem exciting at first but, as time passes, the fun will probably diminish as the novelty fades.

CHAPTER 2

Sex Robots and Violence Against Women

More sex robots, more violence against women?

In the previous chapter, we considered the character of people who purchase and have sex with sex robots. We reached the conclusion it is not true that having sex with a robot is morally reprehensible. Sex robots may also be used by morally appreciable people who are sensitive to the well-being of people dear to them as well as those less fortunate than themselves. But for some, this would still not suffice to justify the use of sex robots in that the robots would still be a threat regardless of the character, motivations, and fantasies of the people who choose to purchase them. We should never forget that the road to Hell is paved with good intentions. So, one can also admit that people who have sex with sex robots do not necessarily harbor the hidden desire to wreak havoc on other people. On the contrary, they may believe that violence is always unacceptable and they may tremble at just the thought of someone deriving pleasure from the suffering of others. They purchase the robots not because they fantasize about possible violence or mistreatment, but simply because they find it an innocent, fun pastime – or simply want to have fun using it with their partner.

Yet, since sex robots are mainly produced with the features of women and their potential purchasers are mainly men, their sale could foster the conviction that women are at the disposal of men or that men have the right to use and exploit them for their pleasure (Gutiu 2016; Sparrow 2017; Richardson 2015, 2016). Having relations with robots, in other words, may seem like an innocent pastime similar to watching a football match at the stadium or spending an evening with friends. It is something that makes us feel better without making other people unhappy. But by dint of having sexual relations with robots however and whenever they prefer, men would start to consider and treat real women as objects. Consequently, they would only form relationships with real women that are marked by violence (Richardson 2015). This means men would first attack the robots and then rape women because by having sex with robots, they would get used disrespecting women (Richardson 2016):

Sex robots however cause harm through positive reinforcement. The user learns violent or degrading practices by physically acting them out with an entity that is indistinguishable from a woman, and is rewarded instantly. During the interaction, at least on a subconscious level, the user will not differentiate between sex with a consenting woman and with an ever-consenting sex robot. This direct form of interaction between the user and sex robot exacerbates the harmful impact caused by pornography (Gutiu 2016, p.11).

In these terms, sex robots would be extremely dangerous first and foremost to women (the predestined victims of the violence), but also to men who would no longer be able to relate to women correctly. They would be dangerous for society overall, as it would be stained by continual violence. For this reason, the production of sex robots should not be permitted as the robots would reinforce power relationships that do not recognize women as legal entities.

Sex robots would, then, have the same consequences for women that pornography does. Pornography contributes to violence against women by "staging" them as people who dream of being raped and enjoy being subjected to violence. They pretend they do not enjoy men's advances, but actually love being brutalized. This is why one can only pleasure a woman by assaulting her (and a man who is unable to assault a woman is not a real man). Sex robots, on the other hand, would support violence against women because they communicate the idea that one can have sex with a woman without asking for her consent. It would be mainly those sex robots who are programmed to refuse any sexual relationship or give the impression they do not enjoy sexual advances at all that would foster this male disposition towards violence against women.

Of course, robots cannot be assaulted because they have no will of their own and so we cannot force them to do things they do not wish to do. Yet their facial expressions, voice, rigidity and reactions may be programmed to have us believe that they do not want to engage in sex or that they can achieve orgasm when they are mistreated. For example, Z-Onedoll sex robots start to groan if they are beaten and the groans increase as the violence intensifies (Devlin 2018). We know that they have no will, but we continually suspend our belief. In reality, any robot could allow one to experience violence against women (even those that seem to enjoy our advances) in that no robot can truly consent to a relationship. Further, the most obliging sex robots would convey the message that women are always available; consequently, it is not really possible to assault or rape them. Indeed, violence presupposes a sexual relationship against someone's will but if women never refuse sex, then they cannot be assaulted. So the risk is that, through the use of robots who are always ready for a relationship, men would lose the ability to accept rejection from a woman seriously. Consequentially, men would enact violence on women thinking that they, too, desire the sexual relationship (Gutiu 2012).

If we take these fears seriously, then there are no other possible paths forward. It is important to call attention to the dangers of sex robots, but this alone is not enough: they must be forbidden, because only by banning their production will it be truly possible to defend women's interests (Rigotti 2019). Describing sex robots as new sex toys would not only be a mistake, but also unfair towards women who may soon face assault from men less and less capable of consensual sexual relations. It is thus necessary to prohibit the use of sex robots before it is too late as if we do not, we will be jointly responsible for the suffering that women experience.

In line with these fears, English sociologist Kathleen Richardson has launched a campaign requesting a ban on sex robots (Richardson 2016).

It has only been a few years since several nonprofit organizations pushed for a moratorium on killer robots (lethal autonomous weapons) (Balistreri 2017, p. 408). The same measures are also being taken today towards robots programmed not to spread death and destruction, but moments of love. According to Richardson (2016), autonomy is a basic value in our liberal societies. For this reason, sex robots are not a private affair but a public matter (Richardson 2016). In this case, the choices of single individuals may harm other people. At first glance, these choices seem to regard only the sphere of personal freedom (the robot serves to achieve pleasure alone or with one's partner), but they would also have the effect of limiting or negating a woman's freedom entirely. According to Richardson, it would be right to intervene to prohibit sex robots. Some people should give up their sexual fancies, and society would derive significant advantages from it: not only would relationships between men and women be more respectful and balanced, but the value and dignity of each person would also be asserted. Prohibiting sex robots would pass on the message that people cannot be reduced to mere objects.

But is it true that sexual violence against robots increases violence against women? That by playing at raping or killing one, we become less sensitive to suffering (making it difficult to resist the temptation to rape and kill in reality)? My generation grew up playing soldiers at war or cowboys and Indians with toy soldiers, wearing a sheriff's star, and even slinging toy pistols and rifles. This experience is common to many males of my generation. As brilliantly recounted by Edoardo Albinati in his latest novel *The Catholic School*, he is amazed that only a few people then became killers (Albinati 2019). Thinking about it gives cause to wonder. As children, we "killed" many times. Now, as adults, we are parents with a respectable job and an immaculate criminal record. After causing so much blood to spill, we now even participate in peace marches and we do not hesitate to express our sincere disdain when we see images of entire families fleeing from war or children who are victims of bombardments.

Maybe this personal-but-collectively-shared experience would suffice in doubting the conviction that what we do in play determines what we will do in reality. Someone may well have become a criminal by profession or been marked by horrendous crimes, but most of us have an irreprehensible (or almost irreprehensible) life. Our character does not seem to carry any trace of our playtime fantasies. This is also true for the games we enjoy playing as we grow older. Even when we have sex, we can give in to "far from conventional" fantasies. During the act, some imagine being their partner's jailer and – while they make love – play at torturing and inflicting violence on them, immobilizing them or limiting their ability to touch and see with ropes, hoods and chains. There is no violence but the parties pretend, one the victim and the other the executioner. Others prefer to play the part of the adult who seduces a female or male adolescent where the adolescent is, of course, the adult partner dressed like a teenager. One partner assumes the role of the dominant, authoritarian adult such as a doctor or a guardian (but even a parent) while the other assumes that of the victim. The result is the simulation of incestuous acts or sexual relations inspired by pedophilia. This is called ageplay and it is a form of BDSM. Finally, others still may be aroused by defecating on their partners who then busy themselves spreading their feces on the body or genitals of both (in a memorable scene from the 1989 film *Crimes and Misdemeanors*, the main character played by Woody Allen complains that while his sister was having sex, she had to undergo the humiliation of being tied up and pooped on by a complete stranger).

We shall put aside the temptation to pass moral judgement on these practices. They are erotic games that many of us may not share. Some may consider them immoral, while others may simply be disgusted. The point is that it does not seem necessary for these fantasies to then transfer to reality; we can get excited in this way without thereby becoming more violent. May we not have the ability to confine our fantasies to a separate world (for example, the privacy of our bedroom or the rooms of an exclusive private club) and only let them out when the time has come to play with other people? Indeed, the nice thing about fantasizing may be precisely this: we are in one place and reality is in another, so we do not then have to answer for our actions because they do not produce any real consequence (Strossen 1995).

However, this is the very thing that is questioned by those convinced that purchasing and having sexual relations with a robot will then necessarily ignite a desire to assault and/or rape in the "customer". For them, it is an incontrovertible truth that this person will become a rapist or killer sooner or later. One who maintains that violence against sex robots will lead to violence against women is saying that violence against sex robots makes violence increasingly exciting (Sparrow 2017). We know very well advertising specifically works on the presupposition that the more the consumer is able to associate the product with gratification, the more sales will rise. So, associating violence against robots with pleasure will result in many people seeking the activity this represents. That means violence against women will rise because far more people will start to fantasize about it and imagine the pleasure it could offer.

The fear that exposure to violence makes people more violent has been hotly disputed by those who have studied the effects of the means of communication (Sparrow 2017). Acts of violence, school brawls, and violent criminal behavior (for example, aggression and theft) happen to be explained by exposure to violent video games. But there is no evidence to prove that drawing pleasure from the portrayal of sex or violence makes individuals more disposed towards violence (Sparrow 2017; Sherry 2007). One who maintains this does not consider that aggressive behaviour is not necessarily maladaptive or socially discouraged (Zagal 2012). Even in the case of people who spend time with violent video games, it is reductive to explain violence as simply learned behaviour because it can result from other influences (Sherry 2001; Ferguson 2007a). Essentially, any attempt to show a direct causal relationship between the content of videogames and violent behaviour should be considered "overly simplistic, largely uncorroborated and ultimately contentious" (Young 2016, p. 30; Markey et al. 2015, pp. 14-15; Huesmann 2010; Ferguson, Kilburn 2010; Ferguson 2007a). And contrary to popular belief, it is not true that the more one plays violent video games, the more one loses sensitivity or empathy towards the suffering of other people (Kühn et al. 2018a, pp. 22-30; Szycik et al. 2017; Kühn et al. 2018b). In this way, the belief that violent video games are a danger can be explained by the moral panic we feel when confronted by new technologies. They frighten us, so we want to demonstrate how bad they are. The same fear we now feel over video games was experienced in the past towards Waltzes, cheap novels, films, rock 'n' roll, comics, television, *Dungeons and Dragons*, and *Harry Potter* (Ferguson & Beaver 2015, p. 242).

The fact that the debate on the effects of 'violent games' has mainly turned its attention to videogames need not be surprising. It is precisely such games that award the player who proves themselves able to carry out actions that would be immoral in the real world, such as stealing, torturing, killing, and raping (Danaher 2017c). The effects of pornography have also been widely analyzed and studied in that for pornography, too, violence against women is often presented as a sign of virility in the men who have sex (Dworkin 1981). But even in this case, research in no way proves univocally that pornography consumption is detrimental (Danaher 2017b). Besides studies underlining a close relationship between pornography consumption and/or greater tolerance towards sexual aggression against women (Wright et al. 2016; MacKinnon 1995; Dworkin 1985; Langton 1993), there is research that finds this relationship is actually minimal or non-existent (Diamond 2009). And as in the case of video games, an increase in the sales of pornographic magazines and movies corresponds with a simultaneous decrease in more violent aggression and murders (Markey et al. 2015). So, there seems to be an inverse relationship between the production and consumption of pornographic material and the amount of sexual violence (Ferguson & Hartley 2009).

Further, pornography could even have a positive effect on the consumer (McElroy 1997). The more pornography is consumed, the lower the risk of violence would be (and, in particular, of sexual violence on non-consenting people). The hypothesis is that the more people are offered instruments and opportunities to satisfy their fantasies, the less likely they will be to enact those fantasies on real people. Pornography would be a sort of safety valve to liberate repressed aggressiveness, permitting the consumer to feel calmer and more relieved. If this theory is correct, then the existence of robots that are able to be assaulted could reduce the amount of sexual violence in society (Sparrow 2017). The more violence that is practiced against the robot, the less violence will be practiced against women. The idea that violence portrayed in pornography or through video games may reduce aggressiveness in people is hotly debated (Gentile 2013), but seems to find some confirmation in research studying the effects of child pornography on paedophiles (Strikwerda 2017). Even if we do not believe that pornography can have a cathartic effect, it can have a liberating one in that it can help bring to light and recount a wide range of sexual habits. In pornography, sexuality that was previously segregated or repressed can be fully recognised in the right to pleasure and in identity (Soble 2002, 2006). Pornography further contributes to transforming our sexual customs, ridiculing sexual hypocrisy, and underlining the importance and value of sexuality. In contrast to sexual repression, pornography could permit people to live their sexuality more spontaneously, with less guilt and perhaps less violence. Finally, the experience of women in pornography is not, as some would like to maintain, universally victimizing. "Pornography carries many other messages other than woman-hating: it advocates sexual adventure, sex outside of marriage, sex for no other reason than pleasure, casual sex, anonymous sex, group sex, voyeuristic sex, illegal sex, public sex. Some of these ideas appeal to women reading or seeing pornography, who may interpret some images as legitimizing their own sense of sexual urgency or desire to be sexually aggressive" (Duggan et al. 1992, p.156).

So are fears towards sex robots excessive? Before advocating for a ban on sex robots, one should probably demonstrate that there is a necessary (or in any case significant) correlation between "play" violence and real violence. Just as someone who kills virtual people in video games is not a serial killer or at least not more likely to become a serial killer, someone who tortures their consenting partner is not a violent person for this reason alone. The same is true of someone who has sex with a robot. They do not seem destined or more likely to become a rapist because of it. Sex toys have also become ubiquitous in the last few years. Until recently, it was mainly men who went to sex shops. But now, an increasing number of women use sex toys for their own pleasure. Despite the general rise in the production and consumption of sex toys (Gentile 2013), there does not seem to have been a concomitant rise in sexual violence. On the contrary (although the public perception may differ), cases of violence against women continue to fall – not only cases of rape and murder, but also domestic violence (Strikwerda 2017). There is more play with tools resembling male and female genitals (and with many other sex toys), but these toys do not seem to have made us worse people who are incapable of building relationships that respect the other person.

This could be further proof that sex robots are not so dangerous. If the fears towards sex robots were indeed justified (and if it were true that having relations with an object representing a gender accustoms us to considering any person belonging to that gender as an object), the last few years would have shown us an indiscriminate increase in violence. If this were true, then violence should be the common denominator in any relationship. In fact, the risk of objectifying the human should actually be greater with sex toys in that vibrators and silicon vaginas identify (and thus objectify) a man with his penis and a woman with her vagina. They reduce man and woman to various body parts: there is no man or woman in their totality, but only through their genitals. So if sex toys have not had negative effects on character, then we should expect even fewer worrying consequences from sex robots. In the case of sex robots, genitals are just a part of the product; neither man nor woman is defined solely by a penis or a vagina.

There is a worrying type of morality behind arguments maintaining that violence against robots can corrupt the character of the people who use them. If we assume that the use of sex robots corrupts people's character, are we not running the risk of condemning or presenting as immoral any artistic portrayal (from literature to theatre, from painting to sculpture) that depicts violence and, by depicting it, calls attention thereto? How can we consider sexbots and their use morally dangerous without at the same time condemning these actual artistic portrayals? Why, for example, should we not also ban De Sade's novels *Justine* and *Juliette*? And are we sure that the depiction of violence (and we can imagine any other type of behavior we consider immoral and unacceptable) is necessarily immoral and cannot have any value? In the case of works with immoral content, the literary experience may allow the reader to suspend their convictions momentarily so as to explore through their imagination the beliefs of the main characters in the narrative. The characters in these stories will probably have moral values unlike ours, but it is precisely this that lets them live an emotionally and cognitively enriching experience in that they will have a chance to consider things from an original perspective – things they may otherwise have never been able to get in touch with or assimilate (Barbero 2012). For example, what drives an individual to inflict violence on another person? Which desires and reasons move them? What pleasure is expected at the moment they decide to torture, attack, or kill another living being? There are also the feelings of a person who undergoes violence at the hands of a man who, at first sight, seemed a reliable person but then proves to be a monster lacking any scruple and capable of the worst barbarity and cruelty. Approaching new situations and experiences as they never have before, the reader has the chance to imagine lives very different from their own and to gain first-hand experience with feelings and passions they may have never felt before.

Of course, making love to robots is not like reading a novel or watching a film. Robots do not tell us stories that enrich our experience (they can be programmed to do so, but sex robots are for other uses). Still, sex robots may be programmed to train people – particularly younger generations – to assume attitudes that are more respectful of women or educate them in socially accepted sexual norms (Danaher 2017b). It makes sense to ask whether sexbots may help us explore aspects of our sexuality that we struggle to discover with a human partner or whether they may allow us to reach new sexual pleasures unlike those we feel when having sex with another human being (or similar to the pleasures we feel with those similar to ourselves, but far more intense and in some ways more pleasant).

In this case, it would be reductive to connect the value of robots only to pleasure. Even if sex with robots were nothing but a simulation of violence, through it we may have experiences that allow us to understand new things about ourselves. After all, we can imagine erotic publications and videos with a pedagogical function as well as the ability to offer alternative sexual experiences. At a basic level, porn provides information about the body and techniques to facilitate its sexual pleasure – something that is sadly often neglected by our society (Strossen 1995). So, sex with robots could have the same function. In any case, it could help less sexually experienced people acquire skills that can help build better relations.

Sex robots designed not only for men

At the moment, most sexbots are produced with women's features (Wosk 2015). But even now, sex robots with a male appearance are available for purchase. That is, the male version of sex robots is not merely a literary invention (I am thinking of Adam, the main character in McEwan's latest novel). After advertising its female sexbots, Absyss Creations presented a male version, Henry, who immediately appeared on the cover of *New York* magazine (Devlin 2018). Sex dolls are not all female shaped and there is no reason to believe that the sales of male versions would be less than their female counterparts. And they, of course, would be purchased by both men and women (Devlin 2018). In the future, we could also have queer robots that do not embody a defined sexual identity or sex robots that are designed to allow the owners to change the identity according to the situation and on the basis of the preferences they feel at any given moment.

This means that alongside robots showing the forms and features of men and women, we could have robots to which it would be difficult to attribute gender. They could have the body of a woman but the genitals and voice of a man, for example, or the body of a man with the genitals, voice, and look of a woman. In other words, they could embody a wide range of customizable variants allowing any person to find their ideal sex robot or the one most appropriate for them. The statement that sex robots are a threat to women and the demand for a permanent international ban in the name of their interests is indicative of a failure to understand that a sexbot is a tool or toy of pleasure not only for men, but also for women. And given that more and more women buy and use sex toys, more and more women will be expected to desire relations with a sex robot. We may also agree that the success of sex robots will depend on several factors. If they have an increasingly accessible cost as well as an attractive look, they may, like sex toys, become a mainstream object.

The motivations leading a woman to purchase a sex robot would be similar to those leading a man to have sexual relations with a robot. Some women may only wish for sex with no strings attached. Indeed, with a robot there is no emotional involvement or responsibilities as it will expect nothing from you (Levy 2008). There will be no living together, weddings, or New Year's Eve parties – nor will there be any need to decide whether to have a child. But who knows if one may, in the future, be able to have a child with a robot. In addition to being a synthetic companion, the robot could be able to carry a pregnancy (given an artificial uterus) or even be the one to reproduce. After all, one can imagine robots with highly advanced technology capable of bringing robotic offspring into the world after selecting the best evolutionary traits and features. This is the scenario described in the *Star Trek: The Next Generation* episode in which Data, an android and second officer of the starship *Enterprise*, decides to have a human-looking artificial offspring. We have already produced robots that are able to design their offspring in turn, using previously gathered information. Maybe the idea of a robot one day being a parent is not so absurd after all (Brodbeck et al. 2015).

For other women, sex robots could be a compulsory choice. They may prefer to have a human partner but are unfortunately unable to find one, or unable to find the right one, and so turn to a sex robot as the next best thing. They need sexual intimacy, like all of us, but fail to find success in the sentimental domain. Perhaps they are not sufficiently attractive or are simply afraid of diving into a relationship. But it may also be that they are too professionally committed, spending so much time at work they cannot dedicate themselves to anything else. They may be successful women whose social position scares or discourages people who are less confident or simply too timid to start courting.

Still, other women will turn to sex robots out of boredom. They want to experience something new, whether alone or with their partner. They may have no difficulty finding a casual partner. Alternatively, they may have been married for years or with a steady companion but the sex is not like it used to be or perhaps is just a memory. And even if the sex continues working and turns out to be very satisfying, some women may have needs that their partner does not meet. Requirements could be different and reflect another sexual desire. The sexbot could even be a solution suggested by therapists, should a man suffer from erectile dysfunction (Cox-Goerge & Bewley 2018).

Finally, individuals may actually be able to have a variety of relationships with a robot. "A robot will be able to provide endless variety in terms of its conversations, its voice, its knowledge and its virtual interests, its personality, and just about every other aspect of its being." (Levy 2008, p. 208). We could, then, change anything as we like – not only height and constitution, but even genitals. In this way, any person could build for themselves the partner they want and indulge in their fantasies to explore their sexuality. And when the robot loses its appeal, it may be replaced by another or one may simply add new functions and parts.

There would be nothing strange in a woman paying to have sex. After all, men pay money to have new sexual experiences (Monto 2001; Levy 2008). So, why might a woman not also be ready to pay to enrich and possibly change her sexual routine? Like a man, she, too, could turn to paid sex (in fact, prostitution is not limited to male consumers as women pay for sex as well). Robots could offer an interesting, valid alternative to prostitution itself. Not only would they be safer (with no risk of attack from a procurer (pimp) on duty or threat of contracting venereal diseases), but they could also be the most economically convenient solution in the long term. Robots would be far more reserved than any casual partner in that they could be programmed to delete their client's information at once and not reveal client fantasies to third parties. But of course, there is still the risk that a robot's memory can be hacked along with the sensitive data used by the company producing the robots to further profit from their sales.

Will sex robots make prostitution disappear?

While women are not the only ones who prostitute themselves, they certainly represent the highest percentage of people working in the sex industry. But should the development of sex robots affect the prostitution market? This would have significant consequences, especially in terms of the lives of women. There are typically two criticisms forwarded against the acceptability of paid sex. One regards violence and the lack of consent, while the other regards objectification and lack of empathy (Danaher, Earp, Sandberg 2017). Prostitution and violence are necessarily linked for the following reasons. First and foremost, prostitution is never fully voluntary because there is always some coercion behind a woman's decision to put her body up for sale. Even when there seems to be no exploitation behind paid sex, the woman finds herself in a situation of necessity because no woman would ever willingly sell her body (Moran 2015, p.408). While some women state that prostitution was a free choice for them, their testimony is not necessarily reliable in that they may not have the courage to reveal the exploitation they suffered (Moran 2015).

Further, prostitution is inseparable from violence because people who prostitute themselves always run the risk of suffering violence at the hands of their clients – and if not the clients, then at the hands those who are exploiting the prostitutes themselves (Thomsen 2015). For many, prostitution seems a morally unacceptable activity because a person's subjectivity is always denied in paid sex. Indeed, the dignity of the person selling their body is never recognized in that they are at the mercy of the client who can dispose of them as he most wishes (Westin 2013). This means that between client and prostitute, there cannot exist a relationship between two subjects because one party (the client) perceives the other (the prostitute) as only a thing. This is why some clients consider sex with a prostitute to be a form of masturbation. They pay a person but for them, the person more object than human – a sexual toy rather than a being with dignity (Richardson 2015).

But this vision of the sex industry has been criticized for its partiality. Prostitution is not necessarily a dangerous activity for women; rather, it has become so due to the stigmatization and social prejudice accompanying it (Moen 2012). Let us start with health risks. Women who prostitute themselves are afflicted with significant psychological problems such as panic attacks, depression, and insomnia. Moreover, the percentage of suicides among prostitutes is six times higher than the average (Farley & Barkan 1998; Bullough & McAnulty 2006). Yet the origin of these problems is not found in the type of activity they practice, but in the disparagement and disgust they receive from other people. So it is not paid sex that makes these women feel bad, but the fact that they are discriminated against and marginalised by a society that disapproves of what they are. There is harm in prostitution, but not because it is a harmful activity per se – not any more so than many other professions (e.g. boxing) that are not forbidden (Earp 2015; Moen 2014). By forbidding prostitution or not fully recognising it, the law hinders prostitutes from obtaining better working conditions (Moen 2014).

The greater the difficulty in renting a flat or room to work in, joining trade unions, negotiating a salary and working hours, and stipulating medical insurance, the higher the risk of suffering violence and mistreatment. Where prostitution is legalized, it is a relatively safe activity in that most commercial transactions occur without violence (Sanders et al. 2009) similar to what happens in non-commercial sexual relations (Danaher, Earp, Sandberg 2017). The traditional vision of commercial sex does not take into account that relations with a prostitute may also be emotionally engaging, just as rich and intimate as long-term monogamous ones (Bernstein 2007). There can be people who pay only to achieve sexual pleasure, but there are also others who seek different things in paid sex such as a person who listens to them or shows interest in them and their life. In exchange for money, they hope to receive the affection and care they do not find in non-commercial relationships because they have difficulty getting in touch with other people or simply because the love in their marriage or relationship has died. The clients of prostitutes are often not in search of mere genital pleasure, "but are rather seeking a 'fiction of intimacy', which may also contemplate a lack of interest in having complete sexual relations" (Casalini 2013, p. 307).

It is, then, understandable that one takes non-commercial relations as a model – but one should acknowledge that such relations are neither intrinsically nor exclusively models of authenticity, intimacy, and reciprocity (Danaher, Earp, Sandberg 2017). People who love each other do not demand money for sex. Yet equally authentic relations based on reciprocal respect may also be built with people who sell their bodies to live. While this may not always happen, we should not be surprised when it does. Do we think that the clients of prostitutes are people without scruples? At times it is so, but at other times they are simply people who imagine they can build truly satisfying relationships simply by paying. They probably know that it is mere fiction, but prefer to deceive themselves and imagine they are in front of a person who loves their company. If people interacting with robots for company are able to suspend their belief and imagine they are dealing with real animals or real people, why should we not be tempted to take fiction as reality when we are faced with a person who smiles at our words and looks at us with affection and love?

Furthermore, it is not clear why paying someone (the prostitute, in this case) for a service without showing any empathy towards them reduces that person to an object. In daily life, we buy services from other people without worrying about them in the least or feeling any form of empathy towards them. Perhaps this morning, we failed to notice that the person who sold us our newspaper or bus ticket was not the same person who did so yesterday. We do not pay attention to them in that we do not stop to chat, for example, or ask them about their families and how they feel or what they are going to do that day. Yet does this fact actually have negative consequences for them? Is this an unacceptable form of instrumentalization? Why would paying a prostitute in order to reach orgasm (but without having any interest in her) amount to a serious lack of respect? Why is this attitude deemed unacceptable because it reduces the other person to an object?

It could be said that it is one thing to sell newspapers or tickets, and quite another to prostitute oneself. Here, it is not an object but one's own body that is sold (Westin 2013). Yet aside from the fact that not even the people prostituting themselves are truly selling themselves (if anything, they concede to the use of certain parts of their bodies for a period of time under certain conditions), there is always some kind of trade of the body behind any type of work in that no profession could ever be carried out without physical commitment. This is more evident in the work of a masseuse or soccer player, but it also applies to those activities requiring the development of more intellectual skills (Moen 2014). Prostitutes work with particular parts of the body (i.e., the genitals), but does this suffice to expose them to either a greater risk of instrumentalization or nothing less than condemned to becoming an object for a client? Those who defend the acceptability of prostitution are skeptical in this regard. First and foremost, prostitutes can negotiate their performance with their clients. As a consequence, clients are only allowed to do certain things. Prostitutes can certainly be forced to do things they do not want to do, but this has nothing to do with prostitution itself – such coercion is sexual violence or slavery. It is not difficult to imagine activities other than prostitution wherein people might receive compensation for the use of their genitals without being thereby instrumentalized or treated as mere objects by others (Nussbaum 1998).

Even if prostitution damaged the person practicing it in some way, the benefits of this activity might more than make up for its costs. For example, young people with no resources or education could easily find a job and invest their free time in other activities. Those in better economic conditions could also derive advantages from prostitution in that they could save their earnings and have money to buy a new washing machine, repair their car, or simply enjoy a vacation. Through prostitution and in exchange for sexual acts, writes Ole Martin Moen, she could receive "something more valuable in return from a larger pool of people. This is a benefit that should be counted" (Moen 2014, p.7). No one wishes to deny that poorer people could have better reasons to prostitute themselves, but this does not mean their choice is not free.

Otherwise, we should conclude that most of us are not free because in order to live (buy a home, send our children to school, etc.), we are forced to work and accept what the market offers us.

My intention is not to take up a position on the acceptability of prostitution. This is a complex matter. What interests me more is showing how selling sex robots could perhaps significantly contribute to reducing the market for prostitution. People who agree to pay for sex today could be tempted by sex robots tomorrow, ultimately preferring them to paid sex. Moreover, there are varied opinions about paid sex. If we think prostitution a fully respectable profession, the development of sex robots could be a large problem given that it can result in a significant reduction in employment opportunities. But if we are convinced that prostitution is always a form of exploitation, we should gladly welcome the production of sexbots in that they could facilitate a reduction of violence against women. Sex robots are not yet so technologically advanced at the moment, and having sexual relations with a robot is not the same as having sex with a human being. But tomorrow, the market could take on technologically improved prototypes that can ensure the consumer sexual gratification similar to that from a sexual relationship with a human being.

And even if sex with a robot were less pleasant and gratifying, we could still have other reasons to prefer them to paid sex. For example, society's attitude towards sex robots could be less critical than its current one towards prostitution. In having sex with a robot, it would no longer be necessary to hide or ensure that one is not caught by others or the police. A minority of people may boast about having paid sexual relations but the majority of clients visiting prostitutes have a different attitude, preferring that others not find out about this form of sexual conduct. There may also be moral reasons for preferring robots to prostitution. Some people believe that it is always wrong to pay a woman for sexual relations. They habitually turn to prostitution but do so in shame, and would not if there were valid alternatives. Others could prefer sex robots to paid sex because they have a companion or a wife and do not want to cheat on them (and also worry over transmitting venereal diseases to them). They could have sex with robots at will and it would not be real infidelity in that these machines are not living beings, but simple artifacts. Indeed, sex robots could be far cheaper than paid sex. There is no need to buy one; one could rent a robot for a brief period, perhaps from a catalogue, and return it several days later. Alternatively, there are always brothels. There are already brothels with sex dolls in operation (Devlin & Lake 2018). Maybe tomorrow, the offerings will become even greater and include sex robots. One may depend on a wide variety of models in that sex robots will not only be programmed with different personalities, but also display different physical features suited to any fantasy. Further, sex robots could expose the client to fewer risks of disease or infections. The risk of contracting venereal diseases can be controlled in sex with a prostitute but with robots, the risk would be far lower. The hygiene and cleaning of sex robots should never be neglected, of course, but they could easily be sterilized.

The question remains whether robots could become a truly desirable object, taking the place of human beings even in sex work. Some harbor profound skepticism regarding the possibility that sex robots could, in the future, become a far more exciting object than paid sex. Others doubt that robots will reduce the demand for paid sex, whatever the pleasure they may be able to provide us with. In any case, we should not assume the attitude we currently hold towards sex robots as a safe reference point for predictions about the future, given that things could change faster than we imagine. Research carried out in recent years (see Chapter 1) shows a growing number of people who can imagine having sexual relations with a robot – from 9% of the sample interviewed in 2013 to more than 40% in 2017. The research is different here and so is the sample, but the data remains eloquent.

Even if sex robots become a widely consumed object, it is also possible that prostitution will not disappear completely or only a very small number of people will no longer turn to paid sex. One thing (sex robots) does not necessarily exclude the other (prostitution). After all, the spread of pornography has not been met with a decline in prostitution. On the contrary, prostitution has grown hand-in-hand with the growth of pornography. Likewise, sex dolls have not affected the sex industry in the least. There is thus no evidence that sex robots will supply artificial sexual substitutes and reduce paid sex (Richardson 2015). It remains possible that tomorrow, there will remain many women prostituting themselves alongside sex robots. It is also possible that with the development of robotics and the increasing mass entrance of technological robots into the field of services and professions, prostitution may become one of the few remaining non-mechanized activities (Danaher 2014).

The success of sex robots will not simply depend on producer ability to bring to the market increasingly intelligent prototypes that can meet expectations along with consumer tastes and preferences. The importance that robots acquire in our society and the value we give to our relations with them will probably be influential, too. The more acceptable that relations between human beings and robots seem (i.e. something normal that we should not hide or feel ashamed about), the more common having sex with a robot will likely become. This is the scenario imagined by Francesco Verso in the 2008 novel *E-Doll*, where androids are built solely for human pleasure. They are almost immortal creatures in that they can independently defend themselves, thanks to billions of nanobots, but they are not marginalized subjects. Instead, they are "people" with whom humans interact. The fortune of sex robots could also be affected by our attitudes towards prostitution and the sex industry. The more that paid sex is collectively perceived as a way to exploit women, the more sex robots will look like a possible, morally acceptable – maybe even preferable – alternative.

While we might legitimately be skeptical about the possible disappearance of prostitution, we should also consider how scientific and technological development may, in the future, open up opportunities we cannot now even imagine. Scenarios that today seem unthinkable, or at least science-fictionesque, could tomorrow become the world that future generations inhabit. The current difficulty in imagining a robot as a partner in a couple's relationship or as a casual companion for a one night stand is understandable. At the moment, sex robots look like an updated version of inflatable dolls. But in the future, sex robots may change and appear to us in a different light. Thanks to advancements in artificial intelligence, they may be able to satisfy our affectionate and psychological needs as well as our sexual desires.

Sex robots as sexual assistants for those with disabilities?

Sexual freedom is increasingly recognised as a basic human right (Appel 2010). Having the possibility to make one's own choices regarding one's own sex life without hindrance from the state or third parties allows one to have greater control over one's needs. Freedom creates access to experiences that are not only pleasing, but may also have positive effects on health[i]. For instance, people with disabilities are often unable to enjoy a satisfying sex life (Forsch-Villaronga & Poulsen 2020). They may not be able to masturbate or form a relationship and have sex due to their specific problems (Appel 2010).

Some do not consider it prostitution, but for others it is still paid sex. The fact remains that there are people performing an activity in favour of those with disabilities that consists in large part of caring for their sexuality. This image is well known to a wide audience by now, thanks to Ben Lewin's 2012 film *The Session* that portrays the relationship between Mark O'Brien, a Californian journalist paralysed by poliomyelitis, and his sexual assistant Cheryl Cohen Green. One of the main supporters of this service in Italy, Max Ulivieri, explains how:

Sexual assistance is a service consisting of having a team of specialists at one's disposal: from psychologists to sexologists to the out-and-out sexual assistant, who lets the person with disabilities get in touch with their sexuality. The way is decided case by case: there are situations in which the person needs to live an experience to be spurred on; in others, there is the necessity to satisfy a substantially physical need. The sexual assistant does not promise to be a knight in shining armour: they just allow one to get in touch with this part of themselves. (Ulivieri 2013)

Turning back to overall examples from other countries too, he recounts how "the sexual assistant meets the person who contacts him/her together with the psychologist and decides what to do and how to do it. There may be caresses and stimulation" (Ulivieri 2013). The draft of the proposal prepared by Ulivieri to legally recognise the sexual assistant as a new professional in Italy does not consider the possibility of penetration, fellatio, and cunnilingus. But in countries that recognise this act (such as Holland, Denmark, Germany, and German and French Switzerland), a complete relationship is not excluded (Quattrini & Ulivieri 2014).

For some, it does not matter whether or not the relationship is complete. It is prostitution in that there is paid sex. Even if a complete sexual relationship is excluded, there is still sex. But for others, sexual assistance (directed exclusively towards people with disabilities) is in no way a form of prostitution. To carry out this profession, one must attend out-of-house training courses on matters ranging from medical, legal, social, sexologic, and ethical. Training "should help the sexual assistant both to get to know the different forms of disability and take into account the specific difficulties [...] according to each case, and to manage the possible emotional involvement" (Casalini 2013, p.303-304). To perform the job of a sex worker, Fabrizio Quattrini explains, special training, psychotherapy, and supervision are not needed. Things are different if one wishes to educate another person in what constitutes a "healthy" sexual response (Quattrini 2014). In this case, the professional must have adequate training in matters of sexual wellbeing, the nature of the disability, and erotic-sexual responses.

Moreover, the primary objective of the sexual assistant is not earnings. Care is mainly given in terms of the relational dimension with the disabled person: they have sexual relations with him (or her) because that is part of their job, too, but sex is not the only thing that is consumed. A trusting relationship is also built, letting the person with disabilities feel desired and loved (Quattrini 2014; Quattrini & Ulivieri 2014). According to De Vries, sexual assistance is precisely the work of caring for a person with disabilities (Casalini 2013). As with any commercial relationship involving the sale of a service, it is ostensibly in the interests of a prostitute to have return clients. But the sexual assistant has a limited, pre-established number of meetings with the disabled person, going from a minimum of five up to a maximum of ten or fifteen sessions (Quattrini & Ulivieri 2014). The time that the sexual assistant dedicates to sex is marginal; the meetings provide for educational, rehabilitative training starting with an informative, practical part on affectivity in order to reach sexuality through the experience of physical contact, massage, and finally "the sensory emotion of masturbation" (Quattrini & Ulivieri 2014, p. 58). Reaching orgasm is only a moment on a path aiming to give the disabled person a greater or new awareness of caressing, embracing, and feeling bodily emotions.

So whilst the sexual relationship is the occupation of any prostitute, the sexual assistant predominantly works on the exercises prior to sexuality – practicing relaxation and re-educational exercises, for example. Physical contact or the chance to hug and touch one another only comes at the end, after work on affectivity, emotion, and corporeality (Casalini 2013). So, the sexual assistant is simply a mediator between the person with disabilities and the possibility of living sexuality happily and without shame. Care for the person's pleasure is only functional to their psycho-physical well-being. Finally, there is one more distinguishing element between a prostitute and a sexual assistant. Whilst the prostitute's work (even in cases in which there is exploitation) does not provide forms of coordination with other people or professional figures, the sexual assistant operates within a triad relationship "which provides for the mediation and supervision of a therapist, a mental health professional, who can be a psychologist or a psychotherapist, mediating the relationship" (Casalini 2013, p.306).

The need to distinguish sexual assistance from prostitution depends on the fact that many people believe paid sex is a form of exploitation (Quattrini 2014). Because one thinks that paid sex is "paying for rape," one feels the need to say that sexual assistance is another thing. But even though attempts to distinguish sexual assistance from prostitution are appreciable and highlight a specific aspect of this profession, they are still not so convincing. First and foremost, who says that prostitutes only think about satisfying the sexual desires of their clients and do not care for other needs? Prostitutes, too, may conceive of their profession as a mission able to give moments of tenderness, authenticity, and sweetness to those less fortunate (Danaher at al. 2017c; Sanders et al. 2005). Some women see their work as an important service that they perform for men who are unable to satisfy their sexual needs alone or build relations that can offer sexual pleasure (Sanders 2005).

Then, there are women who almost exclusively work with disabled people (De Boer 2015). The fact that these women define their job as a "social service" and appreciate their clients' gratitude and acknowledgement shows how reductive it is to conceive of prostitution as a commercial activity in which there can be no room for altruism (Sanders 2005). And while sexual assistance generally only addresses people with clear disabilities, it is not true that prostitution must involve any person at all. People turning to prostitution often have difficulties that hinder having satisfactory sexual relations. Sometimes, difficulty in finding a partner lies behind the choice of paid sex. At other times, there is dissatisfaction with an affectionate relationship (Weitzer 2005). We perhaps cannot call this a disability, but it is still a condition that no longer allows people to have a truly satisfying sex life. By this, I do not mean to say that prostitution carries out a therapeutic function or should be considered a form of sexual assistance. But those who turn to paid sex may expect the same benefits from prostitution as those that people generally expect from sexual assistance (Sanders 2005).

In addition to this, the relationship between a sexual assistant and a disabled "client" is not always mediated by a sexologist who step-by-step determines the path to follow. Both the sexual assistant and the person with disabilities can turn to a therapist or other professionals for help in coordinating and defining their meetings. But they may also not do so. In such cases, Garofalo Geymonat reminds us that there is "a direct relationship between the client and the sexual assistant" (2014). Like prostitution, sexual assistance is mainly practiced by women. This is why some feminists fear this new profession may encourage and promote further exploitation of women (Casalini 2013).

The importance of sexual assistance for people with disabilities is not at issue. But it can, in fact, create another chance to abuse women's bodies in that care is usually considered a female prerogative and it is precarious, poorly paid work (Tronto 1993). It is true that in countries where sexual assistance is legally and professionally recognized, it is work that men practice, too. So, there are male sexual assistants – not only to the advantage of women, but of men as well. However, the request is still mainly from men ("the demand continues to be prevailingly male," writes Casalini 2013) and mostly satisfied by women.

To reduce the risk fostered by sexual assistance for people with disabilities, voluntary work designed for this service could be encouraged (Di Nucci 2017). This would mean that sexual assistance would no longer be paid and the activity would then be presented very differently compared to prostitution. As it would not be a paid job, there could be no more doubts over the altruistic motives of people willing to practice it on the disabled. Sexual assistance would not lose professionalism in that we already have sufficient examples of professionals dedicated and committed to voluntary work. Yet only with robots would women be relieved of care work, in that it would no longer be them but the robots who would take on patient needs. Admitting (but not conceding) that sexual assistance is a form of prostitution and, like prostitution, condemns other women to exploitation, the use of sex robots would have even more significant benefits for females as it would evidently reduce the risk of abuse and violence.

For this reason and contrary to the usual situation, sex robots should arouse the enthusiasm of those who maintain that sex work is not just a job like any other and that prostituting oneself can never be a choice. We may likewise disagree that prostitution is always violence against women. The fact remains that sex robots could reduce the demand for paid sex and this should be good news for those who do not accept prostitution (Royakkers & Van Est 2015).

It is obvious that the preceding considerations apply in this case, too. Just as it is currently difficult to establish whether sex robots may reduce prostitution or make paid sex a far from tempting solution, so we do not know whether sex robots may become a valid alternative to sexual assistance for people with disabilities or other problems. We should also consider the possible risk that, as a consequence of sex robots, people with disabilities may become further isolated from the social context. The person with disabilities could turn to sex robots because they feel lonely and as soon as they have a robot by their side, they stop seeking a human partner (Sullins 2012; Döring & Pöschl 2018). At the moment, the greatest worry is that the sexual assistant, at least in the form providing the mediation of a health official or therapist (sexologist), shifts the sexual desires of the person with disabilities to a medical problem. If sexual assistance essentially performs a therapeutic function, then the sexual desire of the person with disabilities must be considered an illness rather than the expression of a legitimate desire for a satisfying, affectionate life and full inclusion in society (Casalini 2013).

Using robots may lead to different scenarios. There will be people interested only in satisfying their pleasure and, for them, having a relationship with another person will not be so important. They may wish to face their problems with the help of a therapist (psychologist), already have other significant relationships (relatives or friends, for example), or simply not want to have a relationship at that moment. In such cases, the sex robot may present itself as a fine option – even preferable to the services of a sexual assistant. Still, other people may demand something more, feeling a need to satisfy their pleasure and also enjoy human contact with a person willing to listen to them and capable of reciprocating their feelings. In this situation, it is more difficult to imagine how the robot could appear interesting. If this occurs, it means sex robots would have become capable of competing with human beings on equal terms – not just in terms of satisfying sexual pleasure, but also in affectionate, sentimental relations.

Sex robots should be able to not only receive love, but also reciprocate it. They should not only listen to our suffering, but suffer together with us. They should laugh at things that make us smile and be moved when we are moved. This is the only way they can become our ideal partners. If they cannot do so, perhaps simply thinking that they empathize with us will suffice. But it is possible that we will not be able to cope with the company of machines showing sentiments and emotions that they do not actually feel and have. In this case, only a self-aware robot could meet our expectations but we do not yet know whether we will actually be able to make them one day. Regarding robots for the sexual assistance of people with disabilities, it can be further hypothesized that their success will depend on economic cost. If they are cheaper than "regular" sexual assistance or their cost is high but they can still be rented for brief periods, a greater number of people could be interested in their performance.

Further, someone with economic problems who would not like to go without personal contact could alternate between meeting an assistant and having a relationship with a robot. Using the robot may also bring down costs for sexual assistance as, if the sexbot's performance were cheaper than any human one, public health services could cover expenses for the sex robots. In such cases, poor people could have difficulty turning to the services of an assistant. The fact that public health services could pay to let its less well-off citizens enjoy a satisfying sex life would be nothing new. On some countries, sexual assistance is already covered by the state. In others, people with erectile dysfunction can access pharmaceuticals free of charge and the same holds true for women regarding the pill (Appel 2010).

CHAPTER 3

Loving a Robot?

Can one be unfaithful to a sex robot?

For the moment, let us put aside our traditional idea of unfaithfulness where the husband returns home unexpectedly to find his wife in bed with her own lover, or the wife walking downtown with a girlfriend happens to see her husband kiss another woman (or man) in a bar. I do not mean that these things will no longer occur or are destined to disappear. There will still be people caught by their partner flirting with others and/or discovering their partner in the arms of other men or women. But new technologies offer us more and more opportunities to be unfaithful that did not previously exist. We can hardly even imagine them. It is true that in some cases, it may be difficult – not only for the people directly involved, but also for one looking at things from the outside – to establish objectively whether the appearance of unfaithfulness is a matter of betrayal or an innocent game.

In the past, too, a person could get upset and suspect unfaithfulness if he discovered his partner regularly exchanging letters with another person;

the more explicit the letters, the stronger the suspicion became. The same scenarios may present themselves today with new information and communication technologies. The temptation to unlock a partner's mobile phone can be irresistible for some. But by doing so, they may discover their partner's flirty conversations that may be better left unseen. Sometimes, there is not even the need to find explicit messages. It may suffice to discover photos of people in risqué lingerie or in sexy positions, prompting the discoverer to shout, "Cheater!" The accused partner may justify themselves, of course, swearing that they were unaware of ever receiving those messages or photos or assuring that it was just an innocent game and never went further than exchanging texts. But for the other person, this justification may not be enough. The longer these exchanges went on, the more betrayed the other person may feel. In some cases, just one text may be enough to wreck a relationship. Trust in one's companion is lost and any pretext is good enough for an argument. Finally, one realizes the situation has become unbearable and the best thing to do is break up. At other times, the partner may feel reassured by the fact that it was nothing other than messages; disappointment and sadness may remain, but infidelity is another matter. Ultimately, there was no sexual relationship or even an exchange of tenderness and kisses – just shared fantasies.

Yet things become more complicated if we consider other scenarios. Imagine, for example, we find our partner in the act of having sex online. Is this infidelity or should the fact that they are having sex online or at a distance make us far more indulgent towards the sexual extravagance of the person we love? Sex can be had online in different ways. For example, one can get excited just by exchanging texts with another person (sexting). The start of the text conversation is about this and that, but then it turns to far more intimate, personal topics. Finally, one ends up explaining one's fantasies. Some find this exciting, yet orgasm does not necessarily have to be the arrival point for these conversations. One may have many reasons to get in touch with other people online; one can seek love just as much as interesting people and new friends. Sex is certainly one of the many things it is possible to do on the computer. One can also have sex with another person using the telephone or in a video chat. These things are no different from what occurs in an exchange of messages. But on the telephone, it is more difficult to deceive ourselves about the age and gender of the interlocutor. In texting, even a child can present themselves as an adult, a woman, or a man. A very elderly lady can present as a shy girl. Overall, we can never be sure that the other person is who they tell us they are.

Whether behind a computer screen or telephone, someone can assume a different identity and gradually seduce another by saying things about themselves that do not exactly correspond to reality. It is for this reason that meeting a person one got to know from chatting online may turn out to be very disappointing in the end. Video chatting does not present this inconvenience as one has the chance to not only hear the voice, but also see the other person. But even in this case, deception is possible because we can never be sure that what the other person says is true. In any case, through video we can (think that we) know much more about him or her. Even if people cannot have true physical contact in any of these situations, the pleasure of virtual sex may still fulfill some and occasionally prove even more satisfying than what they generally achieve with their partner.

But with new technologies, one can have sex with another person more interactively through a tactile perception of the other person's body (Gomes and Wu 2017). The only things needed are penetrative gadgets for women and penetrable objects for men. Vibrators and "masturbators" (i.e., teledildos) work as genital stimulators and precisely reproduce their anatomy, dimension, functionality, and movements. A man simulates sex with a penetrable object that is able to send a sensory response through a PC or Wi-Fi connection to the vibrator used by the other person on the other end of the screen. At the moment, we are not dealing with technology that can offer participants the same physical experience as one feels when one really makes love with another person. But with scientific and technological developments, the future could afford better instrumentation. The pleasure of a relationship from a distance may become more intense both in terms of the amplification of sexual pleasure and the relation or physical contact. At that point, sexual relations from a distance may be not only a valid alternative or even more tempting than physical sex (Verso 2018).

What is the fairest reaction to a companion who has sex with another person online or at a distance? Is it normal to feel betrayed? Some couples may also have an open relationship and consider it fully acceptable to go out or have sexual relations with other people. For them, the question of unfaithfulness does not present itself. Nothing changes when one's partner has sex with another person in a hotel room or on the computer in that their relationship does not provide for sexual faithfulness (Easton & Hardy 1997). They may even wish to recount and share this experience with their partner or recommend it to the other as something pleasant. But it could be a problem for couples who have promised to be sexually faithful to each other. Here, we have not only messages and photographs – there is more.

The fact that the partner at the other end of the screen can be anonymous (perhaps he has a nickname and his true identity is unknown) is irrelevant in that sexual relations can also be had with perfect strangers. But it is not even necessary for him to be anonymous. It could be a person we know and with whom we started to have online meetings. True, there is no exchange of bodily fluids in online sex because pleasure is reached by way of a conversation or instruments permitting sex from a distance. Yet the sexual activity is undeniable. Although it is remote, it is consciously practiced with another person who in turn writes messages or whispers certain words over the telephone or moves in a certain way. It is not an act of self-eroticism before an image in a magazine or at the thought of a person. It is more: it is a relationship involving the participation of another person in which sexual pleasure is sought. Brett Lunceford observes how no matter what means are used, the point is that online sexual involvement can be far more intimate than any bodily relation. Two people can still become one, but in a way that transcends physical limitations. We must also consider that when it comes in online sex, writing, voice, and especially teledildonics permit extending one's sexual participation to the point of reaching the other person (Lunceford 2009; Faustino 2017). It becomes is as if there were a downright physical relationship with the person located at the other end, who actually cannot touch, embrace, or kiss us in reality even if they wanted to. This may, of course, also apply to sex practiced on the telephone or with messages. But it is mainly teledildonics that allows filling the physical distance existing between two people, as they permit "every participant to alter the other's physical experience in ways that simulate presence" (Lunceford 2009).

It would be more difficult to speak of infidelity if the pleasure reached online were the product of a relation with an interactive image produced by a computer, or an application interacting with us in place of a real person. If masturbation is not infidelity (when we tell our friends that our partner has cheated on us, we do not mean we caught them masturbating), then neither is self-eroticism in front of a computer. The basic ingredient

– the other person – is missing in both cases. So, our partner could still assure us that we are the object of their thoughts and there would be no chance to ascertain their true fantasies. The fact remains that we could legitimately be disgusted or feel some disquiet if these practices became a regular pastime for our partner, or if they preferred to have sexual relations with a computer rather than with us. But if this did not happen, we would perhaps have no reason to complain. On the contrary, the fact that one's companion has sex with a computer (masturbates) may reduce the probability of betrayal.

For the same reasons, perhaps we should not consider sex with a robot infidelity either. Robots are not people, but sex toys that we can use to satisfy our sexual desires alone or in company. As Levy states, a woman who has sexual relations with a robot is no more responsible for cheating on her partner or her husband than the millions of women who regularly use a vibrator (Cheok et al. 2016). But many people seem not to share Levy's point of view. A survey carried out in 2013 revealed that 42% of the interviewees considered having sex with a robot as a betrayal of their partner, while only 31% stated that it would not be (26% could not decide). The percentage drops from 42% to 36% with people under 30, but rises to 52% when the people answering are over 65 years old (*Huffington Post* 2013). Similar results were obtained in the 2017 YouGov survey: 32% of the people interviewed (36% of the women, 29% of the men) saw having sexual relations with a robot as a clear instance of infidelity (YouGov 2017). Further, a study conducted on 848 heterosexual women showed that women may become just as jealous of humanoid robots as of other women (Szczuka and Krämer 2018).

Still, the reactions of the majority of these people hardly seem justified. In the future, sex robots may increasingly resemble human beings – not just physically and psychologically, but also to the touch. One day not so far off, we may become incapable of distinguishing them from our likeness. But it is less probable that they will have self-awareness or be able to make autonomous decisions (aside from the ones they are programmed with, that is). Research on artificial intelligence may make very significant progress, but it will surely be a long time before robots of this type can be built. Until that day, robots will remain objects at our disposal. This is why we cannot cheat on our partner by having sex with them. Can we cheat on our partner with a dildo or any other object? If not, then we cannot cheat on them with a robot either because the robot is not a person.

At the same time, this does not mean our partner cannot feel sorry. They may wonder why we had sex with a robot rather than with them. Are they unable to give us the same pleasure that we reach with the robot? Or they could suspect we have fantasies that we are not brave enough to reveal to them. Perhaps we are thinking of another woman or another man with whom we would like to have sex? We could also spend too much time with the robot and neglect our relationship, or give attention and passion to the robot that we cannot enjoy in interactions with our partner. Think of people who today spend a lot of time caring for their cars; the same could happen tomorrow with robots. This would not be betrayal, but it would still be reprehensible behavior because caring for this object could take a lot of time away from more important things. The robot cannot reciprocate our feelings, but someone could manage to love it and grow as fond of it as if it were a real person – actually appreciating and relishing its attention. The fact that someone fell in love with their robot could become a valid reason to break up a marriage or a relationship. But this will depend on the intensity of this passion and the consequences that the presence of this inconvenient third party has for the relationship and life of the people involved. It is one thing to seek the company of a robot now and again, only to relieve our boredom and solitude; it is another to insist that the robot be present at family birthday dinners or parties with friends. The ability to share this choice with one's partner will be crucial. Having sexual relations with a robot against the wishes of a partner or unbeknownst to them may not actually be infidelity, but it could still weaken the trust between partners.

Things become more complicated when the sexbot resembles a famous celebrity (Micklethwaite 2017). We already have synthetic replicas of porn stars on the market. Tomorrow, we may find robotic copies of cinema stars, national team soccer players, and television actresses for sale. Famous people have fans who go crazy at the idea of possessing their clothes and obsessively seek to approach, touch and kiss them. Any one of these fans would surely find the chance to have sex with a replica of their idol irresistible. There are intellectual property rights so authorization would be necessary, but famous people are often happy to link their image to commercial products. They derive significant compensation from this link, which also increases their fame. There may be some worry about a moral or opportunistic order to begin, but these reservations will likely gradually disappear over time and with greater acceptance of robots.

Some famous people may refuse to act as a model for the mass production of sex robots, but others will gladly agree. Having sex with a replica robot is not like having sex with a person. We do not truly embrace and hug this person – just their synthetic copy. So if there must be another person in order to be unfaithful, the problem does not present itself here either; the relationship is not with a person, but a robot. The robot may certainly resemble a person yet despite the resemblance, it is only a machine. Still, it is true that there is not only a robot here in that there is also a reference to a particular person. Their body, their voice, and perhaps even their genitals can be reproduced. So, is it not understandable if the fact that one's companion had sex with this type of robot is perceived as a sort of betrayal? This is not a matter of debate due to a physical relationship with another person (the robot is not a person!). It is because we do not simply want our partner to not betray us; we also want to be the sole object of their sexual desire. We would like them to wish to have sex with us alone, not with others. In this case, the person we love is fantasizing about another person.

However, the fact that one chooses to purchase a robot with a physical resemblance to a specific person does not necessarily prove they would like to have sex with that person. Nor does it mean they would rather have sex with the original than its synthetic copy, but only settled for the robot because they could not get the original. One might purchase that particular robot because it is cheaper than other models (it could be on sale), or because it displays features of an industrial, technological design that other robots do not have. Likewise, it could be the only sex robot model still available at a nearby outlet. It could be light and easy to disassemble, and hence more easily transportable. Others may choose the model deliberately, but still clearly distinguish between imagination and reality. While they are in the arms of the robot, they could fantasize about the person it represents (by caressing its arms and breasts or brushing its hips and kissing it, imagining they are really touching and kissing that person). At the same time, they may not have the desire to have sex with this person for a variety of reasons. Perhaps they deeply love their partner and would never do anything to hurt or disappoint them, or they sincerely believe in the value of sexual faithfulness and intend to maintain their moral integrity. They enjoy playing with their fantasies, but do not cross the border of imagination. For some, it may not suffice that their partner does not consummate the betrayal; they may already consider it unacceptable that the person they love has given into this type of fantasy. Betrayal, then, is not consummated with another person but is still imagined. Yet it can be considered perfectly normal to have erotic, sexual fantasies regarding other people, even join in this fun with one's partner, and not have any difficulty in sharing one's fantasies with them. Both could experience these moments as occasions to reinforce their relationship.

Loving a robot?

Things would be different if sex robots became the object of our love. Indeed, having a sexual relationship with them in this situation would be blatant betrayal. Someone could lose their mind over their robot and prefer an exclusive relationship with it rather than splitting their time between the robot and their partner. In this case, there would be nothing strange about their partner feeling betrayed. Of course, this scenario would be a true revolution in that for the first time in the history of humanity, human beings could have a machine as a partner. There are people who prefer the company of an animal to that of a human being. For some people, machines are more important than animals and human beings. With the advent of increasingly intelligent robots similar to human beings (humanoids), having a relationship with a machine could become normal. We must not think that there will then no longer be any relations among human beings. But alongside human-to-human relationships, we could find more and more between human beings and robots.

There may be different opinions about the features a robot should have in order to become an object of not only desire, but also love. This is precisely what we will cover in the following pages. That is, we will try to understand what may make robots possible candidates for our affection and our feelings. But we can already state that we have never had a partner of this type. The robot may accumulate experience, for instance, but it cannot age. It can malfunction or sustain damage, but it can always be repaired. Its memory may stop working, be switched off, or deactivated, but it cannot die because its data can be saved to another electronic device. It can have abilities that a human being could never achieve, not even with the most assiduous, prolonged, physical and intellectual activity. Its memory can successfully accumulate an impressive mass of data and information of which a person cannot even dream. A robot can also work beyond the most extreme limits of human endurance and never tire because, even after intense physical activity, it does not need to stop and rest or sleep. It can make us lunch, tidy the house, clean the bathroom, wash the floor, load the washing machine, hang up and iron the laundry, change light bulbs, fix things, and still want to look after us and listen to us. It would, of course, be a tireless lover. It would never have a migraine or be exhausted, and would always be able to satisfy any desire we might have. All this may look like a remote, almost science-fictionesque scenario. Yet scientific development could make it a reality tomorrow. We still do not have robots capable of substituting for humans in every activity, but the advance of robots is so unstoppable that anything could happen.

The theme of love between human beings and virtual characters, children, and robots resembling humans has long been explored by television series, documentaries, and films. For example, in *Lars and the Real Girl* (2007), Lars has a so-called real doll as his companion. In *Ex Machina* (2013) and *Humans* (2015), the relationship between human beings and increasingly sentient robots lies at the center of the story. Intimate relations between humans and robots no longer exist only in science fiction. Now, they have even crossed the threshold of academia to become a research field called Lovotics (Cheok et al. 2017). But in the future, will it really be possible to love a robot? According to Levy (2008), we will not have to wait so many years before we see the emergence of loving relations between a human being and a robot. It will probably take much more time before a robot capable of self-determination is built, but a robot could have us lose our heads even before it has free will. What it does will be chosen and programmed by others in the planning phase of production, but it may be able to learn autonomously on the basis of its experience.

For Levy (2008), relations with a robot will be able to accommodate anything that is important to us in a traditional loving relationship. We often think it is important to share certain interests with our partner, for example. One may also have this relatively easily in relations with a robot. The robot can be programmed with specific interests and thus be able to share our enthusiasm at an opera première, or when a film with that actor we love so much comes out at the cinema. It will jump for joy when we want to go to Saint Petersburg again. But it will be bored to death over anything that does not interest us and react with disapproval when someone says or does things that we dislike. There will be no arguments over where to go on holidays or to celebrate New Year's festivities. This is a result we cannot hope to reach even with the person we love most because although we may share interests and passions, get on perfectly well, and can count on a wonderful connection, there are always differences that cannot be overcome – and it may be better that way. Otherwise, a relationship can start to bore us after a while. We might wish for something else because it would be as though we are facing not another person, but a mirror of ourselves. Everything would be predictable, which creates the risk of monotony.

But if that is what concerns us about a sentimental relationship with a robot, the problem does not exist. We can simply see to it that the robot has a personality of its own, with different tastes, interests, and preferences from ours. Of course, people change over time and become different – but we can predict this too. The robot could be programmed to adapt to new habits or, after some time, we could simply reset its personality. What goes for its personality and its character goes for everything else too. For instance, do we want the robot to be attractive and have a sexy attitude? Its physical appearance can be modelled according to our preferences for attractiveness and sensuality. Robots have no smell but with new technologies, we could incorporate artificial smells – the ones we find most pleasing and sexually stimulating – into them.

But we do not simply want to have a marvelous (personality- and body-wise) object to love. We also want our companion to love us for who we are. We want them to miss us when we are absent and then, when we return home or see them, to be happy to see us and embrace us. In Levy's opinion, this is not a problem either. Indeed, it is not necessary for the robot to actually love us. The impression that it loves us will suffice, namely, by reacting promptly and most appropriately to our expectations. For this to happen, robots will need to be far more intelligent in that such robots must not only look human, feel human, talk like humans, and react like humans: they must also be able to think, or at least to simulate thinking, at a human level. They should have and should be able to express their own (artificial) emotions, moods, and personalities, and they should recognize and understand the social cues that we exhibit, thereby enabling them to measure the strengths of our emotions, to detect our moods, and to appreciate our personalities. They should be able to make meaningful eye contact with us and understand the significance of our body language. (Levy 2008, p.119)

At the moment, we are not able to produce robots which can interact with us in a truly satisfying manner. We do not yet have the necessary technology but one day, they could become a true object of our love.

The first step to such an end will be having deep, detailed knowledge of the human physique, physiology, and emotions in order to reproduce them in a robot. Then, it will be a matter of endowing the robot with software capable of analyzing input from the human partner and generating adequate moods and behavior. Finally, the robot's affective system needs to be emotionally appealing (at least for the human beings who have continuous relations with them). That is, the robot's reactions should change in accordance with the affection they receive and the reactions their partners would like to receive in exchange. Robots do not produce hormones such as oxytocin, serotonin, and endorphins. But they could read human facial expressions, tone of voice, body temperature, and blood pressure to react accordingly. The expression of emotions could be further emphasized through movement to approach or distance, vibrations, and sounds. At that point, they could seem like people who really are in love.

But the robot will not be self-aware or able to reciprocate our love. So, how can we think that a robot would be capable of making us fall in love? We do not only want to have the impression that our partner loves us. We want to be truly loved (or that is the conception of romantic love, at least) (Nyholm and Frank 2017). Otherwise, an actor playing the part of falling in love would suffice for us to feel loved. Hiring such an actor full time would become an expense, albeit probably still cheaper than a therapist. We could pay them by the hour, when we feel lonely or want to feel that we are in a relationship. Yet even if the actor is particularly talented, it will not be true love. At most, we could describe it as a portrayal of love, as we would like love to be experienced between two people who are fond of each other, or as we have seen it enacted at the cinema or in the theatre.

But there would be no love. It would seem a simulation of love at best – not an exemplification of it.

Maybe these obvious considerations show that it is impossible to fall in love with a robot (Nyholm and Frank 2017). As long as there are no robots intelligent enough to be self-aware or reciprocate our feelings, the idea that there may be authentic love between a human being and a robot seems pure science fiction. We can imagine it and find it related in a novel or film. We can even passionately follow, from start to finish, the difficulties the robot and the human must overcome to crown their love (there will certainly be those who are against their love, but all obstacles will be done away with in the end). In reality, a human being will never be able to feel love for a robot

Any person who loves another would doubtlessly want that love reciprocated. We do not mean only in terms of receiving gifts such as a bunch of flowers or an expensive item on a birthday. We also want to know that he or she really feels something – that we are always in his or her thoughts, and his or her behavior is the mirror of our behavior. So in order for us to fall in love, is it necessary that our love be reciprocated? Not exactly. As adolescents or even later, we can go completely crazy for successful tennis champions or rock stars. We may buy all the magazines discussing or publishing photographs of them. We cover the walls of our room with posters of them. We go to their concerts and refuse to miss a tournament in the hopes of catching their eye. Some even reach the point of writing ardent letters and awaiting a reply. But these figures are so far away and they cannot even imagine that we exist. These feelings we experience towards them do not touch their heart.

And when we are older, it may also happen that our love is not reciprocated. How many times does one end up falling in love with the "wrong" person? Of course, there is nothing wrong with the fact that a person does not reciprocate our feelings; it is only wrong in the sense that this is not the person from whom we can expect love. We may think of nothing but them. We blush and stammer when we meet them while they are completely indifferent and have no reason to suspect that they have been in our thoughts for days on end, almost like an obsession. Romantic stories generally portray the outcome of such situations as the suicide of the one in love (think of Goethe's *The Sorrows of Young Werther* or Flaubert's

Madame Bovary). In reality, men whose feelings are not reciprocated may even reach the point of murder.

We can also continue to love people who used to love us but no longer do so. This love can be so gripping and pathological that we are willing to accept any compromise just so we do not lose it. Instead of dealing with reality, we imagine that things have not changed. We accept not only betrayal, but also mistreatment and violence. Moreover, the fact that we can continue sincerely, passionately loving one who is no longer there further confirms that love does not need to be exchanged. How many people continue loving their partner long after his or her death (Sisto 2018)? While falling in love with a robot may seem strange, it does not at all seem impossible. If we can even love people who do not reciprocate our feelings or have stopped loving us, why should it be impossible to fall in love with robots?

Robots cannot love us but ultimately, this aspect does not seem so indispensable. What may constitute a problem is not so much the fact that they are not alive (after all, establishing what is alive and what is not remains a very complicated philosophical question). Rather, the problem is that they are not sentient. But with the advance of robotic technology and artificial intelligence, a robot may be able to react adequately to our stimuli despite its lack of both feelings and the ability to empathize. And are we really sure that robots have no feelings? We are used to thinking that feelings are something inside the subject. But if robots are able to emote us or make us fall in love and even get jealous (as Adam does in McEwan's latest novel), how can we think that they have no feelings? Feelings are not subjective, but relational in character (Dumouchel & Damiano 2017). From what we know regarding the use of robots for the care and assistance of the elderly, the idea that a person may one day fall for a robot does not seem absurd at all. Our ability to empathize with robots is already a reality (Coecklebergh 2010). It is now increasingly evident that we can also grow fond of robots, attributing human features and feelings to them.

Those with doubts, Levy (2008) rightly observes, should think of what happened just a few years ago with the virtual animal Tamagotchi. The Tamagotchi's time is up. New generations do not even know what it is, but the phenomenon of getting attached to virtual realities can still be observed. Now, the inanimate object that has won over the hearts of millions of Japanese people is Rinko, the star of LovePlus (a videogame for Nintendo DS) who allows players to simulate a sentimental relationship with another person (Liberati 2017). Something of this kind was later developed in the West too, with applications such as *My virtual girlfriend*. For the player, the objective remains the same: conquering one of the available virtual girls through different forms of courtship. According to Phillip Galbraith, the relationship with a LovePlus character is a real one; "this type of relationship is immediately gratifying and it is always available" (Silvestri 2014).

Virtual entities such as Tamagotchi or the LovePlus "girls" are not real. But someone who interacts with them daily experiences the same fascination and infatuation that children generally feel with their favorite toys. Children have a strong bond with their teddy bears, soft toys, and dolls. Many of them treat these toys as if they were living beings (Beschorner & Krause 2017). The same phenomenon has been observed studying the use of robots in the field of care and assistance for the elderly. These people are aware that the robot is not a human being but for them, it is something other than a simple machine (Sharkey & Sharkey 2012). They get to the point of thinking they have a responsibility towards the robot, treating it like a friend or (less frequently) like a partner (Coeckelbergh 2010). So, it is not hard to imagine what might happen tomorrow with increasingly intelligent, technological robots who may be indistinguishable from human beings or humanoids and appear to have the same feelings as we do. If the robot converses and behaves in the same way as a human being or if it "can produce the same (or greater) experienced levels of companionship, satisfaction and emotional comfort (than) a lover can" (Nyholm & Frank 2017, p. 223), then we may not resist the temptation to treat them like humans and love them.

It is still early to establish whether we could actually have loving relationships with robots in the future. This will require determining how far new technologies will be able to produce robots that are both physically and psychologically similar to human beings. Indeed, one can imagine how the more humanoid the robots of the future are (the more they resemble human beings physically and psychologically), the easier it will be for us to perceive them as humans and treat them as our peers. It is true that animating these robots in a way that does not cause our disgust may turn out to be far more complicated than it could seem at first glance. It is the problem of the Uncanny Valley (or Zone). Indeed, it has been observed that the more human robots seem and the more similarly they behave – albeit not in exactly the same manner – the easier it is for human beings to feel a strong sense of repugnance and disgust towards them.

A possible explanation for this phenomenon, documented for the first time by Masahiro Mori (2012), is that we are dealing with an adaptive reaction selected by evolution as useful in keeping us at a distance from ill or strange people. After all, the same repulsion is felt towards corpses and zombies. At the same time, another hypothesis has also been put forward. In this one, observing the unnatural movements of humanoid robots would force us to face our mortality as the perturbation would generate the same defenses that are activated with thoughts of death (McDorman 2005). According to Blay Whitby, the "Uncanny Valley is an important reason why it is highly unlikely that *physical* robots will be adopted as artificial sexual companions by those with mainstream sexual preferences, at least for the immediate future" (Whitby 2012, p.236). However, Mori's theory requires further study. Moreover, the repulsion and perturbance observed in the Uncanny Valley problem fails if the robots cannot be distinguished from human beings. So the Uncanny Valley does not seem to be an insurmountable obstacle for a relationship between a human being and a smart humanoid robot of the future (Carpenter 2017).

We could also fall in love with robots who only vaguely resemble human beings or are totally unlike us physically. Ultimately, even human beings in the future will have more and more mechanical and technological parts. So, the fact that robots do not resemble us does not have to be a problem. What seems most important for a loving relationship is that the robots are able to appreciate our uniqueness. In other words, they should be capable of showing love for what we are – not only for our qualities and our particular inclinations, but also for our defects and failings (Nyholm & Frank 2017). This does not mean that human beings could not fall in love with far less sophisticated robots lacking the minimal ability to relate to us as individuals with personal experience and history. But it may be easier for us to become intimate with a robot if it is able to learn from experience and keep a memory of our relationship. Building a machine that is capable of learning from experience (machine learning) is possible. It would be more complicated, but it still does not seem impossible to program the robot to assign a special value to the person with whom it interacts. Some philosophers doubt that robots may ever have this capacity, while others are far more optimistic regarding the possible developments of artificial intelligence. Regardless of who is right, Nyholm and Frank (2017) observe that in order to possess this capacity, "the artificial intelligence in such a robot would need to be of a very sophisticated sort. We understand this as being hard to achieve, though not necessarily something that is in principle impossible" (p. 233).

Furthermore, it is foreseeable that our attitude towards robots and the possibility of falling in love with them will partly depend on how they are perceived. In fact, the more they are accepted socially, the easier it will be to have relations with them. Otherwise, one might be ashamed to be seen in public with a robot or to feel a sensation of love for it. For some, this will not be a problem. But most people will worry about what society thinks of their sexual inclinations. And if relationships with robots are considered the expressions of dispositions that are morally debatable (such as asocial or misogynistic tendencies) or capable of fostering sexual violence towards women, love for a robot could be seen as a serious form of sexual deviance. Someone who loves a robot could even be subjected to medical treatment. People would still fall in love with robots. But they would do so in hiding out of fear of being judged and, especially, punished.

Could robots become persons?

At the moment, sex robots are nothing but new-generation sex toys. Compared to traditional sex toys, they go beyond representing just a part of the body to include the entire individual. They are technologically more advanced than silicon dolls and they have a low level of artificial intelligence, which allows them to interact more intriguingly with the people who purchase them. They not only express some basic emotions, but also respond to the questions they receive much as a virtual assistant would. Yet the hypothesis we are discussing is that, with scientific and technological development, they may eventually become more and more similar to human beings. Robots may become far more interactive than the models currently on the market, with fleshy skin and the ability to both recognize and express emotions, with particularly developed sensory perception, with refined language abilities, with the capacity to learn from experience and – far from insignificant – with multiple personalities (Eskens 2017).

It is difficult to foresee what will happen over the coming decades in the field of robotics. How far will scientific and technological development advance the building of robots with not only sex appeal, but the ability to become the object of our love? It is impossible to make predictions about how satisfying these relationships will be or whether they will prove a true alternative to relationships with human beings. For some, robots will never substitute for human beings in sexual relations let alone affective relationships. Still, others are more optimistic. It will take time, but sooner or later we may have the technology that permits us to make robots just as sexually attractive and interesting as human beings.

For now, we are dealing with mere objects that have no moral relevance. They are not self-aware and they do not even feel sensations. Whatever we decide to do, we cannot cause them any suffering. We can use them in the way we prefer or deem necessary. We do not have the moral duties towards them that we have towards people. But if the more optimistic predictions on the development of robotics and artificial intelligence are correct, our perception of robots could change. Robots will be able to look us in the eye and show sincere, deep affection for us. We will think that we really are special people to them. At that point, we will be able to build affectionate relationships with them and even fall in love with them. Love between a human being and a robot seems very bizarre today, but it could become normal tomorrow. We want to be loved. For most, it is important that those who love us have a sincere interest and are not just doing so for convenience or to gain something in return. A robot's affection could strike us as precisely the sort of disinterested love that we are seeking, because a machine cannot have ulterior motives or betray our trust.

At that point, will we still consider robots to be things or will they become people? The answer is easy if we hypothesize robots as being self-aware with feelings. In this case, they will doubtlessly have full moral relevance. It is true that they would not be human beings, but that matter is not relevant: what counts is that they can suffer and we can damage them. We cannot, of course, rule out the possibility that tomorrow, we will be able to produce technologically advanced robots who are self-aware (conscious of themselves) and sentient. But for the moment, this scenario seems remote – almost science fiction. Things would be different if robots only give us the impression that they are conscious of themselves as subjects existing in space and time (self-aware), feeling sensations such as love and joy but also suffering. They would lack self-awareness in that they would not have a mental life or even the minimum sensitivity to pain or pleasure. Yet we would still think they are aware of what is going on. Their reactions could be truly taken for sentiments and emotions. They could show surprise by raising their eyebrows and dropping their jaw. By raising their upper lip and lowering their eyebrows, they could show disgust. Pulling back their cheeks and raising their lips would be a sign of happiness.

But even though they seem to feel emotions, they would have none of those features considered necessary so as to enjoy full moral relevance. We have different opinions on what and which individual deserves moral relevance. Some think that only human beings fully deserve it, for example, and they believe it is right to treat humans as people from the moment of conception (George & Tollefsen 2008). If we only attribute moral relevance to human beings, we cannot consider robots persons because robots are clearly not human. This position, which attributes moral value only to human beings and so assumes an anthropocentric point of view (Midgley 1979), is rightly deemed debatable because it denies moral relevance to animals for no justifiable reason. Animals are not human beings, in fact, but they can suffer. They can not only suffer, but also be rational and self-aware (Singer 1993). One may distinguish between more or less rational animals and there are also animals that are not rational at all. But rationality is not a human prerogative in that there is no ontological difference between humans and animals (Rachels 1990). Even if we assume that one must be rational and sentient to be a person with moral standing, robots (who can only give the impression that they are rational and sentient) may not have moral relevance because they lack either feature.

So would it be correct to conclude that robots cannot be persons? For the moment, we have no reason to treat them as persons in that robots today are nothing but a piece of technology with no intrinsic moral value. Of course, machines may have moral value as well because what happens to machines can have very significant consequences for human beings. We are not overly interested in an airplane exploding *per se* – it is just a machine, in the end – but the explosion of an airplane may cause the death of a great number of people. This is why we think it is important to check the functioning of the computers on board, the engines, the landing gears, the fuselage, the wings, the tail, etc. Not doing so would be irresponsible as well as a criminal liability for the very simple reason that the safety and lives of passengers are at stake.

Yet this does not mean that this is the reason why we consider them people. Machines are objects and for this reason, we can treat them as we wish. But is it not possible that, with the production of robots who are increasingly technological, interactive, intelligent, and both physically and psychologically similar to human beings, our approach to them may radically change? Over the last decades, our attitude towards animals has certainly changed. Today, more and more people think it is wrong to cause them unjustified suffering. Many of us continue to eat meat, but we only buy products that come from farms respecting the wellbeing and integrity of animals. Other people go even further and prefer to follow a vegetarian or vegan diet because they consider any form of animal exploitation unacceptable, regardless of whatever advantages we may obtain. Could what has happened with animals over recent decades also happen in the future with robots who are not sentient, but give the impression that they are? Will we feel the need to treat them differently from how we treat them today?

If we consider things rationally, we have no reason to expect that our attitude to robots may change (at least in the near future). Obviously, if we hypothesize that robots will have the same abilities as us, things are different. But until that day, robots do not seem to deserve any consideration. They are things no different from a desk, table, chair, or lamp. When we caress a dog and cuddle it or take it for a walk in the park, we make it happy. But we can also make its life hell by tying it to a chain only a few centimeters long and making it sleep in the cold, depriving it of food for days and days, and beating it bloody each time it does not obey us. We can do the same things to a robot, but without the same consequences. We can neglect it and forget to recharge it or take it to be repaired. We can even mistreat it, kick it, or hit it violently with a club or any other object. But we cannot make it suffer.

Yet human beings do not only consider things rationally: imagination plays a central role in our lives. We saw this before, too, when we discussed the possibility of having sexual relations with robots. Through the robot, we can imagine we have a partner and are making love with it. We can imagine wreaking violence on a non-consenting person. We can also imagine having a relationship with a famous person or a person who we do not know but with whom we would, for instance, very much like to have a loving relationship. The fact that we imagine these things does not necessarily mean that we would like to do them, nor does it mean we are then more likely to put them into actual practice. Imagination is part of our humanity and not something that inevitably corrupts or destroys our moral sensitivity. Above all, we may, through our imagination, take a break from reality, building fantasy worlds where we can dream of being people different from who we really are and performing things that we would never do otherwise.

Here, we have mainly taken sexual fantasies into consideration. But we might also imagine ourselves as the main character in an adventure novel, saving the princess locked in the tower or fighting to free oppressed peoples. Our fantasies do not have preset contents and limits; they can freely fly. And with the help of the imagination, we can also bridge the distance separating us from the lives of others. We can put ourselves in their shoes, look at things through their eyes, and even feel what it means to be them. The importance of this imaginative process for moral life is evident. Indeed, by cultivating the imagination, we are in a better condition to perceive aspects of reality that other people, less able to sympathize, do not recognize (Balistreri 2010). The rational viewpoint is clear. Robots have no moral value because they are things, so we have no reason (apart from interest) to take care of them. We have no duty towards them. We can do what we like with them. But with the help of the imagination, could we reach different conclusions? Robots are not people, but could we imagine that they are?

In *Westworld* (2016), a television series broadcast by HBO about a theme park where visitors relive the experience of the Wild West, the main characters are robots – or rather humanoids, because they are indistinguishable from humans. Tourists can do whatever they like with these humanoids. They can have a love affair, but also kill, torture, and wreak havoc on them. There are no legal consequences because robots are machines and objects have no rights. But not everyone has fun. Some visitors show empathy towards the humanoids and the spectators do not seem to be indifferent to what is going on either. As Coeckelbergh (2010) explains, what counts here is not what is in a robot's mind; it is what we feel when we interact with a robot. So for Coeckelbergh, sympathizing with a robot need not rely on some level of robot awareness or feelings. Instead, the ability to put oneself in its shoes and imagine its emotions suffices.

But this is a matter that we humans have no difficulty in doing. First and foremost, we have a natural tendency to anthropomorphize objects (that is, we give them human features) and we do this with robots, too (Dumouchel & Damiano 2017). We can know that they simulate human behavior. But even though we are rationally aware that they cannot feel any emotion, we are able to attribute a complex mental life and many of our own abilities to them (Turkle 1984). This is already occurring today with robots that do not yet possess high levels of intelligence and are still very unlike human beings in the physical aspect. What will happen when there are robots that are (at least almost) indistinguishable from human beings? Will it not become even easier to confuse them with human beings and treat them like people? Is it not possible that we will, in the future, tend to attribute more and more moral relevance to robots (and consequently treat them like downright 'people') because it will seem to us that what is done to them is being done to another human being? Robots cannot suffer but once they are both physically and psychologically similar to human beings, we may imagine that they can. There will be a machine in front of us, but we may have a different perception (Danaher 2019; Gunkel 2018).

The moral relevance of robots and the perception that they are persons would just be an illusion. We attribute moral relevance to robots and consider them persons only because we confuse their appearance with their true nature. They are not human beings. It is we who wish to believe they are, but they remain machines that behave in a certain way because they are programmed to respond to stimuli they receive from outside. If we consider things rationally, they cannot be persons. They are objects as intelligent as we want to imagine, but a product of ours, nonetheless. But in the future, will reason still be able to prevail over imagination? Or will it be forced to give in before increasingly intelligent humanoid robots? In this case, intelligent robots could acquire the same relevance as we have[ii] and become persons.[iii]

We could, of course, continue to ask whether it might not be better to avoid designing robots able to make us believe they are people just like us, in the light of what may happen (the progressive inclusion of robots into our moral community as subjects with full rights and the same moral standing as us). Joanna Bryson does not doubt that robots should be our slaves. Here, she does not mean to justify slavery. She simply warns us of the risk of human beings perceiving robots as their peers. Robots are mere tools and it is important for them to be perceived as objects: we can use them to extend our abilities and to increase our efficiency, analogously with the way we have used other things in the past. But precisely because robots are tools, Bryson believes they do not deserve the moral relevance we may be tempted to attribute to them should they one day become perfectly similar to human beings. Considering them something more than simple objects would just be a waste of time (Bryson 2010).

According to Bryson (2010), the more that robots are treated as humans, the more that humans are destined to lose their own value and seem more and more like things. This is why, she concludes, putting humanoid robots on the market would be a big mistake. The advantages of robotics are not under discussion as the problem is not robots, but our inability to interact with a robot without making it seem to be a living being. The problem is us. We do not always consider things rationally. So, the solution is simple: we must not build robots that can lead us into temptation – that is, robots we can take for individuals deserving moral consideration (Bryson 2010). Robots should not even have a human appearance, if that helps, because this would further reduce the risk of perceiving and treating them like people. We could leave some mechanical parts exposed, plan them with integrated screens, or allow them to be disassembled and usable for other functions. We can do what we want but the important thing is that they do not resemble humans.

Bryson's considerations start from the hypothesis that with the development of increasingly humanoid, intelligent robots, we would no longer be able to consider them as things and we could attribute personality to them. There is a real possibility of this happening, as we previously explained. But it remains to be seen whether, as Bryson suggests, it is better to perceive robots as slaves – that is, as simple objects at our disposal with which we can do what we like – rather than as persons who deserve respect. Bryson is right in that with the production of robots who are increasingly intelligent and similar to human beings, we may be tempted to choose a robot rather than a peer of ours as a partner. This is why the consequence may be a reduction in human relations and, even more dangerously, in the ability to build a relationship with another individual. The other is unique by dint of their identity and character. How can we develop the ability to deal with the other and their diversity if we get used to building our affective relations with robots who are always our mirror?

The risk is that humans become less and less pleasant, interesting, and reliable. We may interact with people with whom it is necessary for work (if there is still work to be found) or institutionally (to decide how to organize a company), but who cannot give us what we can obtain from humanoid robots. The scenario that can most easily be imagined is a future of extreme solitude where our desire for the other could easily be filled by intelligent machines which are not aware, but still seem to be human beings. And we must also add to this the risk that robots be increasingly perceived as entities with the same moral value as human beings, and so deserving to be treated as we treat our peers. If they are persons too, why should human beings count more? Imagine having to choose between a human being and a robot. Who do we want to save? In this case, would it still be wrong to prefer the humanoid to the human being or can one easily imagine situations in which it would be morally acceptable – if not obligatory – to give priority to the robot over the human being?

Say we find ourselves in a burning skyscraper and can only save one person. The possible choice is between our neighbor, who lives on our floor, and the robot with whom we share our life and are in love. Rationally, the answer is simple. We should save the human being. Indeed, our neighbor may suffer. The neighbor also has a clear perception of danger and what may happen to them if we do not help. But the robot is unaware of what is happening. Moreover, it cannot suffer. While we can always substitute the humanoid and replace it with a new machine, the human being is unique and cannot be replaced as if it were an object. However, are we sure that we will see things in this way tomorrow? Independently of what we now perceive and what seems rationally right to us, is it not possible that saving the humanoid's life will seem understandable tomorrow? We may not often find ourselves in a burning building, but the problem of choice will always present itself – not least when we have to decide how to distribute public resources. Should we only deal with the health and well-being of human beings, for example, or should we also think of facilities and interventions necessary for a robot's repair or cognitive development? And if we want to introduce a basic income guarantee, should we only consider human beings – or will robots, too, have the right to some form of income support? Robots would not know what to do with a basic income guarantee, but their owners could use it to look after their robots and improve their functions.

According to Bryson, these considerations suffice to show that it would be a mistake to continue trying to make robots increasingly resemble human beings. The more the robots become similar to humans, the more we lose value and moral relevance. Yet the advantages we may obtain from producing and using humanoid robots should also be considered. First and foremost, we should consider the fact that the more similar to human beings the humanoids become, the more they may be successfully used in areas of thus far purely human competence. Think, for example, of the advantages that humanoids would bring to the field of health. More specifically, think of caring for, accompanying, and assisting the elderly, those with disabilities, and the ill. In interacting with robots that resemble human beings, these people could feel less lonely. They might build affective relationships with robots far more easily.

But once robots have a human appearance, it may also become far easier to utilize them as nurses caring for children when parents are absent. Children might also be able to get used to less human-looking machines. With a human-looking robot, however, they may feel safer and more protected. Adults, too, may prefer to have a humanoid robot going round the house and cleaning up or making lunch, replying to our video call to book a dentist appointment, or welcoming us at a hotel reception or at a bank entrance and checking our documents. Further, one should also consider the advantages that may be derived by people who, due to disability, do not enjoy satisfying sexual, affective relationships with other human beings. Their health can also benefit in that there would no longer be doubts that sexual, affective relationships with other people positively affect our well-being. But humanoid robots may provide a significant resource for any person. Even those with a partner could use them as sex toys, whether to have fun alone or to rekindle relations with their companion. Furthermore, introducing them could reduce or do away with prostitution.

It is not true that once humanoid robots are put up for sale, it will become increasingly difficult to continue having relationships with human beings. Not only are relationships with humanoids compatible with relationships with human beings, but relationships with robots could also facilitate or support relationships with our peers. So it is true that, as we said previously, robots could help people who are lonely or suffer from social anxiety to (re)gain confidence in themselves and other people. Even without taking these scenarios into account, just the presence of a robot could provide the pretext to go out and possibly meet other people. Someone who owns a sex doll might talk about going out with other people who love the same type of company, for example. Then, there is the annual get-together for those who live with dolls, which is another chance to socialize (Devlin 2018). The fact that robots might acquire moral relevance because they would seem more and more like people does not create any danger of dehumanisation (but we could also feel wounded pride because we would lose our uniqueness).

In reality, humans would continue to have the same value and the same rights. It is true that we may end up before completely new situations in which a human being would run the risk of being sacrificed or neglected for the benefit of a humanoid. But in this case, it would still not be correct to say that human beings would be reduced to mere "things" as they would not lose their relevance. Simply put, androids would be perceived as persons, *too*. But if robots became people, our interaction with them would have to be completely reconsidered. We might still have the right to buy a robot and use it in our homes for housework, as a babysitter, or in industry. But would it be a crime to buy one only to enact violence on it, abuse it sexually, and treat it as a mere object? And in this case, would using violence on a robot not be the mark of a moral deficit? Here, we would indeed have "just" a robot – but one who is socially recognised as a person.

Conclusions

As long as we cannot produce robots intelligent enough to be capable of being self-aware – and to be able to reciprocate our feelings – the idea that there may be love between a human being and a robot seems like science fiction. But are we sure that the robot's behavior, its emotional reactions, its attitude of care towards the human being, and its ability to be close to us in moments of sadness will not be perceived as sincere? In this case, we would have less difficulty in thinking there could be love between a person and a robot. One of the two is not a human being, but what does it matter in the end?

It is true that we tend to associate the idea of love with the condition and mood experienced by two people who feel made for each other. We seek in the other person a half-destined part to complete what we feel is missing in ourselves. Indeed, what is Eros if not the desire to reify one's love (Plato, ca. 370 B.C.E./2008)? But when it comes to love between a human being and a robot, there would be an evident asymmetry. The asymmetry lies in the fact that the robot would be made for the human being (indeed, the robot is produced with a program and a certain look because we desire to fall in love with it), while the human being would not be made for the robot (Nyholm & Frank 2017).

Yet we should not think that love can only arise with robots we have designed in detail, physically and psychologically, at the moment of purchase. We might also fall in love with humanoids that other people have designed (not necessarily to be sex robots; they could also be robots for company or entertainment). In a more futuristic scenario, robots could be perfectly integrated into our society and we would be able to meet them as we generally meet other people, such as "in a bar, or on a dating app like Tinder" (Nyholm & Frank 2017, p. 229). These robots could also be different in physical appearance, personality and values, and some robots may even be the right partners for some people. But in other cases, there will be no spark between humans and robots.

But according to Nyholm & Frank (2017), a problem would remain in that robots would not experience what we do when the person we love reciprocates our feelings. Love, they write, is not just feeling and passion. Love is something more; it is a judgment, a decision, the result of a continuous choice. This is an ability that will not be within a robot's grasp, at least for the moment (Nyholm & Frank 2017). But it is debatable whether love, as Nyholm and Frank maintain, really is the result of a conscious choice that we can reason out dispassionately. It could be argued that love is actually a passion. Like all passions, it cannot be controlled. We cannot decide rationally when and whether to fall in love with a person by evaluating, for example, their family, their qualities, their economic situation, their profession, the city in which they live, and their friends. Nor can we stop loving on command when our love is not reciprocated, or when it is reciprocated but still causes us to suffer. For this reason, love may become similar to other forms of dependence. We may be aware that the substance (a drug, an alcoholic drink, etc.) or the person we desire is not good for us, but we still continue to seek it. Behavior observations, neurochemical reactions, and neuroimaging all provide ample evidence that love is (or at least can be) an addiction similar to the way in which chronic behavior characteristic of the search for drugs can be called an addiction; "in some instances, 'treatment' of love could be justified or even desirable" (Earp et al. 2017, p. 79).

So, it is true that robots only act on the basis of the programming they have received. They cannot do anything that their programmers have not previously decided on (they may eventually have this ability but for now, it is not possible). Yet our choices, too, are determined by factors that we do not control (Danaher 2019). This applies not just to love, but to anything we do. Our objectives, the important things that give sense to our existence and to which we dedicate our time and energy, are not chosen rationally. They are the product of our upbringing, our culture, our genetic code, and the relationships we have with the people we meet on our path. No one wishes to deny the possibility that love can gradually wear itself out in that the person we love can always leave us or fall in love with another person. But if this happens, it is not because they have thought it over better. It is because they have stopped loving us. Or they may still love us, but love for themselves prevails over their love for us. There is no such risk with robots. The love of a robot will not fail even when we pay them less attention or neglect them because we are in love with another. We might go further and decide to switch the robot off, but it is not programmed to stop loving us.

But would we not perhaps be missing something if, when we fall in love with a person or a humanoid, we were certain that their love for us would last forever? And are we sure that, when we have fallen in love with a person, we really think of the possibility of love ending and how we may one day love each other no longer? Perhaps we do know, but how present is this awareness when we love? And might we not feel a little less loved if our partner were always aware of this possibility and their choices were always conditioned by an awareness that even the finest things can have a beginning and an end?

Yet if we also want the robot to stop loving us, we can program it. For example, we could ensure that it would react with disapproval or disappointment when we stop interacting with it as a sincere lover would. We can also ensure that, in consequence, it no longer takes the minimal interest in us and stops appearing kind, affectionate, or caring. The robot may not leave our dwelling, but it could still deactivate itself. At the moment in which we are no longer able to love it sincerely, the robot could first stop interacting with us lovingly and then, once its love for us has totally run out, cease to work. Most people are unlikely to ever be interested in this option because robots would then lose part of their commercial appeal. But other people may choose it so as to be more motivated to behave in a certain way. We could, for example, program the robot to love us more when we do physical exercise, or love us less when we smoke too much. Some might choose it to make their relationship with the robot more similar to relations with humans.

Only the coming years will tell us how similar sex robots can become to human beings and the ways this similarity will affect our perception. We will learn whether they can be companions with whom we can also build affectionate relationships, whether we can fall in love with them, and how far love for robots will condition or even modify our relationships with other people – and even animals. This is not a matter we have discussed, but sex robots can also resemble animals. Some may consider it taboo, but bestiality is not necessarily offensive or dehumanising for the human being participating in the relationship (Soble 2002). The question regarding animals is more complicated as it is indeed true that sex with animals is not necessarily cruel (Dekkers 1994). However, it is often violent and can even involve the animal's death (Singer 2001). Can we be sure that an animal is consenting and how can we establish this? But here, we are simply talking about animal sexbots allowing someone to live out their own fantasies without causing any harm.

Further, it remains difficult to predict whether sex robots may change our experience of sexual pleasure. At least for the moment, sex robots are objects that permit both women and men to reproduce the penetrative experience. This means we are dealing with instruments of pleasure that are presented as a much more technological extension of the vibrators and vaginas sold as sex toys today. Sex toys reproduce the anatomy, the dimension and the function of the genitals, but robots promise a pleasing experience with the reproduction of a male and female body. But for robots, too, the genitals are identified as the only place of physical pleasure. Attention to the robot's physical appearance and psychological reactions exclusively serves to make it easier for the user to reach pleasure in the zone of the genitals (in other words, to reach orgasm from penetrating the robot or from being penetrated by a robot). Non-penetrative acts are not suppressed; one can do what one prefers with a robot as there are no limits to the extent of our imagination. But neither are they encouraged or imagined (Faustino 2017). So, the model of coitus as an exclusively penetrative act – one that seems to merge with a teleological, reproductive vision of sexual meeting – is reinforced. The sexual act is presented as instrumental, by nature, for conception. Penetration of the vagina by the penis follows a precise trajectory in which ejaculation is the natural, legitimate point of arrival for the sexual act.

So, robots appear to be a revolutionary tool for sexual pleasure in that they promise the same experience of pleasure as we have with human beings. Yet, at the same time, they seem tied to an essentialist conception of sexuality (to a particular model of sexuality) that still endures, despite the fact that reproduction is increasingly distanced from sex. Even with the advent of sex robots, the three basic moments of the sexual act continue to be preparation for penetration, penetration itself, and finally orgasm. A reproductive function may no longer be attributed to coitus, but the model of a sexual relationship aiming at reproduction is still the dominant one – presented as the only way to build intimacy or closeness between people who love each other.

But sex robots could be useful in overcoming the dominant model of sexuality because even with the use of other technologies, they may permit new physical, sensory experiences and new forms of sexual interaction. We no longer expect sex toys to represent our genitals. Maybe if robots were less human-like, they could give us more pleasure. New forms – such as a velvet or silk body, sensors and mixed genitalia, tentacles instead of arms" and then a design so that pleasure can be reached in all senses – must be explored (Devlin 2018, p. 266). These new forms may help reduce the risk of objectification (Devlin 2018). One could also try to combine robotics with other technologies. For example, sex robots might be connected to one another by computers. In this way, what we do to the robot could be transmitted to another person (through their robot) and felt as a physical experience by them. This would mean our interaction with the robots turns into physical perception for the other person; in turn, the person interacts with a robot connected to a computer or carries instrumentation that captures our signals at the genital level.

In any case, we would be dealing with the evolution of teledildonics in that people could, through robotics, be involved in a sexual experience from a distance. The experience would be with another person and with the whole body, but the sexual experience would be separated from contact and not require physical presence. As Maria João Faustino stresses, "destabilizing the association of sexual interaction with skin-on-skin contact...[means] teledildonics represents another opportunity to recreate and diversify sexual scripts. This opportunity can be taken to promote more fluid and nuanced, rather than orgasm-focused, experiences of sexual interaction" (Faustino 2017, p. 254). That is, the absence of skin-to-skin contact may contribute to questioning an essentialist, heterosexual model of sexuality and the idea that pleasure is linked only to penetration.

Sex robots could make it easier to explore new relational configurations. New technologies may allow us to have sexual relations from a distance with not just one but several individuals, and in ways that we cannot even imagine now. They could allow for exploration of sexual fantasies that do not correspond exactly to one gender identity (male and female) and can derive from an oscillating, ambiguous, and dynamic sexual orientation. So far, robots on the market have had a defined (mainly female) gender. Tomorrow, it may be possible to vary their identity according to our preferences through the purchase of modular parts or add-ons. A robot that looks like a woman could have male genitals, for example; in reality, a transgender converter already exists as Harmony can have a penis rather than a vagina (Fosch-Villaronga and Poulsen 2020). A robot with a man's physical appearance could have female genitals. Of course, all possible combinations cannot be predicted – they could be more varied and left to individual sexual fantasies. Moreover, the participation of women in processes for producing and planning robots may serve to counter gender prejudice and stereotypes (Danaher 2018). Still, this may not suffice as we would also have to "create better contexts for the marketing and use of sex robots. This would require greater 'consciousness raising' around the problems of gendered harassment and inequality, and a greater sensitivity to the representational harms that could be implicated by this technology" (Fosch-Villaronga and Poulsen 2020, p. 15).

What does the future hold for us? And how will robots evolve? Will we be able to produce robots who are increasingly intelligent and capable of being considered real human beings? We cannot yet give an answer to these questions. It is still too early to predict the feelings we will have from relations with sex robots, along with whether our interactions with them will be gratifying or fail to meet our expectations. But starting from previous considerations, it is possible to draw some conclusions. First and foremost, our reservations towards sex robots are understandable; we are ultimately dealing with new technologies that we are not yet used to. Yet looking further ahead, our worries seem largely unjustified. One of the most recurrent objections to sex robots is that only people who are immature and unable to have appropriate relations with other people will have an interest in and desire for sexual relations with a robot. Behind the choice to purchase a sex robot, we expect to find diffidence at best and fear of contact with others – and more likely, an aversion and lack of respect for one's peers. The desire to have a robot at our side would be nothing other than the hope and aspiration to find a submissive, compliant, obedient partner who is always available to satisfy any request of ours or tolerate any mistreatment. There would no longer be any need for foreplay or boring conversations. There would be no duties, obligations, or embarrassing matters. There would be no need to talk. "This sure sounds like a perfect arrangement," writes Hauskeller (2014).

But we have shown that it is a mistake to necessarily link the choice to purchase a sex robot or have relations with it to negative character dispositions only. It is true that sex robots can be bought by sexist, misogynistic people. But in other cases, we are just dealing with the desire to have a more exciting sex toy than a vibrator or plasticized vagina, to find something fun and original for sex games with one's partner, or to temporarily fill a companion's absence or distance (Marrone 2018). Not even the fact that a person has sexual relations with a sex robot programmed to refuse – that is, to say "no" to any sexual advance or relationship – necessarily indicates the involvement of a despicable individual. One may have the desire to simulate violence with a robot, but we cannot conclude from this that the person doing so wants to abuse other people or that they will sooner or later grow tired of robots and go on to attack human beings. Fantasies, play, imagination, and fiction are not regions of reality. Just as one who "kills" virtual people in video games is not necessarily a serial killer, one who tortures his or her consenting partner is not violent solely for this reason. So, someone who has sexual relations with a robot is not destined to be a rapist. Likewise, there is no reason to think that the trade and use of sex robots will contribute to promoting violence towards women. On the contrary, they may reduce prostitution and sexual exploitation.

But open questions remain. For instance, sex practiced on a child or adolescent robot is not actual violence against a human being. But it is difficult to accept the idea that a person who desires to buy and have sexual relations with such robots is not actually a pedophile and does not have other perversions. Maybe here we are simply confusing "fantasy" or the "play" of fiction with reality. After all, why should killing a person in role play or in a videogame be morally acceptable while rape against a child robot or in a videogame should be terrible and the mark of a reprehensible character? This is only a hypothesis and before drawing conclusions, we need further research. We have also seen how with the production of humanoids who are increasingly intelligent and similar to human beings, it could eventually become easier and easier to feel affection for them. Could we even fall in love? And if we one day lose our heads for humanoids, will we still think of them as only machines? We have tried to reflect on these questions, too, but it will be for future generations to establish whether robots can really become our companions and acquire the moral relevance they lack today.

---

[i] This idea is supported by the World Health Organisation's specific sexual health programmes and by the declaration of Sexual Rights written by the World Association for Sexual Health.

[ii] Besides, this is the very conclusion we should draw if we accept the Turing test conditions. If what an automaton or robot does may seem the work of a human being, then the automaton must be declared human and we can attribute 'moral' relevance to it. It will not be truly human, but if it is able to interact with us adequately and with its environment, then it must be considered - and treated - like one of us.

[iii] However, if, as one may imagine, they had a memory that could be extracted and implanted into another machine, their destruction would seem less serious to us and in any case not comparable to murder. But, even assuming that in more serious cases of violence, the robot can be brought back to life by transferring its memories into another machine, we could still have the impression that we have lost something important. After the transfer of its memory, we would be facing the same robot (after all, there is psychological connection and continuity, Derek Parfit, *Reason and Persons*, Oxford: Oxford University Press, 1984), but something could still change for us, because we would no longer have the original, but its reproduction.

BIBLIOGRAPHY

Abel, Gene G., Becker Judith V., Cunningham-Rathner Jerry. "Complications, Consent and Cognitions in Sex Between Children and Adults." *International Journal of Law and Psychiatry* 7 (1984): 89-103.

Abel, Gene G., Gore David K., Holland C. L., Camp Nancy, Becker Judith V., Rathner Jerry. "The Measurement of the Cognitive Distortions of Child Molesters." *Annals of Sex Research* 2 (1989): 135–153.

Adshade, Marina. "Sexbot-Induced Social Change: An Economic Perspective." In *Robot Sex. Social and Ethical Implications*, edited by John Danaher, Neil McArthur, 289-300. Cambridge (Massachusetts): The MIT Press, 2017.

——. "Robot Sex Could Power Up Marriage." Interviewed by T. Fletcher, *University of British Columbia News*, April 17, 2018, https://news.ubc.ca/2018/04/17/robot-sex-could-power-up-marriage/.

Albinati, Edoardo. *The Catholic School*. Translated by Antony Shugaar, New York: Ferrar Straus & Giroux, 2019.

American Psychiatric Association. *Diagnostic and Statistical Manual of Mental Disorders – DSM 5* (5ª ed.). Washington, DC: American Psychiatric Association, 2013.

Anderson, Kenneth, Waxman Matthew. *Law and Ethics for Autonomous Weapon Systems: Why a Ban Won't Work and How the Laws of War Can*, Hoover Institution, Jean Perkins Task Force on National Security and Law Essay Series, 2013.

Andrews, Donald Arthur, Bonta James. *The Psychology of Criminal Conduct* (5ª ed.). Cincinnati, OH: Anderson, 2010.

Appel, Jacob M. "Sex Rights for the Disabled?" *Journal of Medical Ethics* 36, no. 3 (2010): 152-154.

Arkowitz, Steve, Vess James. "An Evaluation of the Bumby RAPE and MOLESTscales as Measures of Cognitive Distortions with Civilly Committed Sexual Offenders." *Sexual Abuse: A Journal of Research and Treatment* 15 (2003): 237–249.

Asimov, Isaac. "Runaround". *Astounding Science Fiction*, (1942): 94-103.

———. *I, Robot*. New York: Gnome Press, 1950.

———. *Robots and Empire*. Garden City, NY: Doubleday, 1985.

Atkinson, Robert D. "'It's Going to Kill Us' and Other Mhyths about the Future of Artificial Intelligence." *NCSSS Journal* (2016): 8-11.

Babchishin, Kelly M., Hanson R. Karl, Herman Chantal A. "The Characteristics of Online Sex Offenders: A Meta-Analysis." *Sexual Abuse: A Journal of Research and Treatment* 23, no. 1 (2011): 92–123.

Bader, Michael. *Arousal. The Secret Logic of Sexual Fantasies*. London: Virgins Books, 2003.

Balistreri, Maurizio. *Etica e romanzi*. Firenze: Le Lettere, 2010.

———. *Il futuro della riproduzione umana*. Roma: Fandango, 2016.

———. "Robot killer. La rivoluzione robotica nella guerra e le questioni morali." *Ethics & Politics* 19, no. 2 (2017): 405-430.

———. " 'Uccidere' e 'stuprare' bambini ai videogiochi: considerazioni morali." *Lessico di etica pubblica* 1 (2018): 73-81.

Balistreri, Maurizio, Zara Georgia. "Il valore etico della ricerca scientifica in psicologia." In *Lo psicologo tra essere e fare. Problemi di deontologia applicata alla professione psicologica*, edited by Georgia Zara, Michele Presutti, Eugenio Calvi, 174-197. Italia: PubliEdit, 2016.

Balon, Richard, ed. *Practical Guide to Paraphilia and Paraphilic disorders*. Switzerland: Springer International Publishing, 2016.

Bamonte, Tom. "Autonomous Vehicles: Drivers for Change." *Roads and Bridges* (summer 2013), 5-10.

Banks, Marian R., Banks William A.. "The Effects of Animal-Assisted Therapy on Loneliness in an Elderly Population in Long-Term Care Facilities." *J Gerontol a Biol Sci Med Sci* (2002), 57A: M428-M432.

———. "The Effects of Group and Individual Animal Assisted Therapy on Loneliness in Residents of Long-term Care Facilities." *Anthrozoos* 18 (2005): 396–408.

Banks, Marian R., Banks William A., Willoughby Lisa M. "Animal-Assisted Therapy and Loneliness in Nursing Homes: Use of Robotic Versus Living Dogs". *Journal of the American Medical Directors Association* 9, no. 3 (2008): 173-177.

Barbaree, Howard E. "Denial and Minimization Among Sex Offenders: Assessment and Treatment Outcome." *Forum on Corrections Research* 3 (1991): 30-33.

Barbero, Carola. *La biblioteca delle emozioni. Leggere romanzi per capire la nostra vita emotiva.* Milano: Ponte delle Grazie, 2012.

———. "Pornografia." In *Manifesto per un nuovo femminismo*, edited by M.G. Turri, 139-149. Milano: Mimesis, 2013.

Barnao, Charlie. *Le prostitute vi precederanno. Inchiesta sul sesso a pagamento.* Soveria Mannelli (CZ): Rubettino, 2016.

Bartneck Christoph, Kanda Takayuki, Mubin Omar, Al Mahmud Abdullah. "Does the Design of a Robot Influence Its Animacy and Perceived Intelligence?" *Int. J. Soc. Robot* 1 (2009): 195-204.

Beard, Jack M. "Autonomous Weapons and Human Responsibilities." *Georgentown Journal of International Law* (2014): 617-681.

Beck, Aaron T. "Beyond Belief: A Theory of Modes, Personality, and Psychopathology." In *Frontiers of Cognitive Therapy*, edited by Paul M. Salkovskis, 1–25. New York: Guilford Press, 1996.

Behrendt, Marc. "Reflections on Moral Challenges Posed by a Therapeutic Childlike Sexbot." In *Love and Sex with Robots. Third International Conference LSR 2017*, edited by Adrian David Cheok, David Levy, 96-113. Cham (Switzerland): Springer, 2017.

Beier, Klaus Michael, Ahlers Christoph J., Goecker David, Neutze Janina, Mundt Ingrid A., Hupp Elena, Schaefer Gerard A. "Can Pedophiles Be Reached for Primary Prevention of 28 Child Sexual abuse? First results of the Berlin Prevention Project Dunkelfeld (PPD)." *The Journal of Forensic Psychiatry & Psychology* 20, no. 6 (2009): 851-867.

Beier, Klaus Michael, Grundmann Dorit, Kuhle Laura Franziska, Scherner Gerold, Konrad Anna, Amelung Till. "The German Dunkelfeld Project: a Pilot Study to Prevent Child Sexual Abuse and the Use of Child Abusive Images." *The Journal of Sexual Medicine* 12 (2015): 529-542.

Bemelmans, Roger, Gelderblom Gert Jan, Jonker Pieter P., de Witte Luc. "The Potential of Socially Assistive Robotics in Care for Elderly. A Systematic Review." In *Human-Robot personal relationships*, edited by Maarten H. Lamers, Fons J. Verbeek. Third International Conference, HRPR 2010 Leiden, The Netherlands, June 23-24, 2010 Revised Selected Papers (pp. 83-89). Dordrecht, The Netherlands: Springer, 2011.

Bemelmans, Roger, Gelderblom Gert Jan, Jonker Pieter P., de Witte Luc. "Socially Assistive Robots in Elderly Care: A Systematic Review into Effects and Effectiveness." *J. Am. Med. Dir. Ass*, 13 (2012): 114-120.

Benbouriche, Massil, Renaud Patrice, Pelletier Jean François, De Loor Pierre. "Applications de la réalité virtuelle en psychiatrie légale: la perspective de l'autorégulation comme cadre théorique [Self-regulation and virtual reality in forensic psychiatry: An emphasis on theoretical underpinnings." *Encephale*, 42 (2016): 540-546.

Bendel, Oliver. "Sex Robots from the Perspective of Machine Ethics." In *Love and Sex with Robots. Second International Conference, LSR 2016, London, UK, December 19-20*, edited by Adrian David Cheok, Kate Devlin, David Levy, 17-26. Cham: Springer, 2017.

Bennett, Casey Carroll, Sabanovic S., Piatt Jennifer, Nagata Shinichi. "A Robot a Day Keeps the Blues Away." In *IEEE International Conference on Health Informatics* (ICHI Park City, Utah, USA), 2017: 536-540.

Beres, David. "Perception, Imagination, and Reality." *International Journal of Psychoanalysis*, 41 (1960): 327-334.

Berg, Olsen M. "Sex Robot Will Say No If 'It's Not in the Mood'." *Metro News*, June 19, 2018. https://metro.co.uk/2018/06/19/sex-robot-will-say-no-not-mood-7642591/.

Berlin, Fred S., Sawyer Denise. "Potential Consequences of Accessing Child Pornography Over the Internet and Who Is Accessing It. Sexual Addiction & Compulsivity." *The Journal of Treatment & Prevention* 19, no 1–2 (2012): 30–40.

Bernstein, Elizabeth. *Temporarily Yours: Intimacy, Anthenticity, and the Commerce of Sex*. Chicago: University of Chicago Press, 2007.

Beschorner, Thomas, Krause Florian. "Dolores and Robot Sex: Fragments of Non-antropocentric Ethics." In *Love and Sex with Robots. Third International Conference, LSR 2017*, edited by Adrian David Cheok, David Levy, *128*-137. Cham (Switzerland): Springer, 2017.

Bilger, Burkhard. "Auto Correct." *The New Yorker*, November 25, 2013.

Birnbaum, Gurit E. "Beyond the Borders of Reality: Attachment Orientations and Sexual Fantasies." *Personal Relationships*, 14 (2007): 321-342.

———. "Bound to Interact: The Divergent Goals and Complex Interplay of Attachment and Sex Within Romantic Relationships." *Journal of Social and Personal Relationships*, 27 (2010): 245-252.

Birnbaum, Gurit E., Mikulincer Mario, Gillath Omri. "In and Out of a Daydream: Attachment Orientations, Daily Couple Interactions, and Sexual Fantasies." *Personality and Social Psychology Bulletin*, 37 (2011): 1398-1410.

Birnbaum, Gurit E., Svitelman Neta, Bar-Shalom Adi, Porat Omer. "The Thin Line Between Reality and Imagination: Attachment Orientations and the Effects of Relationship Threats on Sexual Fantasies." *Personality and Social Psychology Bulletin*, 34 (2008): 1185-1199.

Bitna, Kim, Benekos Peter J., Merlo Alida V. "Sex Offender Recidivism Revisited: Review of Recent Meta-Analyses on the Effects of Sex Offender Treatment." *Trauma, Violence, & Abuse*, 17 (2016): 105-117.

Blackford, Russell. "Robots and Reality: a Reply to Robert Sparrow." *Ethics and Information Technology* 14, no. 1 (2011): 41-51.

Blizard, Deborah. "The Next Evolution: The Constitutive Human-Doll Relationship as Companion Species." In *Love and Sex with Robots. Third International Conference, LSR 2017*, edited by Adrian David Cheok, David Levy, 114-127. Cham (Switzerland): Springer, 2017.

Bonta, James, Andrews Donald Arthur. *The Psychology of Criminal Conduct* (6ª ed.). New York: Routledge, 2017.

Borenstein, Jason, Pearson Yvette. "Robot Caregivers: Harbingers of Expanded Freedom for All?" *Ethics Inf Technol* 12 (2010): 277-288.

Bostrom, Nick. "Ethical Issues in Advanced Artificial Intelligence." In *Cognitive, Emotive and Ethical Aspects of Decision Making in Humans and in Artificial Intelligence*, edited by George Eric Lasker, Wendell Wallach, Iva Smit et al., vol. 2, 12-17. Int. Institute of Advanced Studies in Systems Research and Cybernetics, 2003.

———. *Superintelligence. Paths, Dangers, Strategies*. Oxford: Oxford University Press, 2014.

Brodbeck, Luzius, Hauser Simon, Lida Fumiya. "Morphological Evolution of Physical Robots Through Model-Free Phenotype Development." *PLOS ONE* (2015): 1-17.

Brody, Stuart. "The Relative Health Benefits of Different Sexual Activities." *Journal of Sexual Medicine* 7 (2010): 1336-1361.

Broekens, Joost, Heerink Marcel, Rosendal Henk. "Assistive Social Robots in Elderly Care: A Review." *Gerontechnology* 8 (2009): 94-103.

Brooks, Rodney. *Flesh and Machines: How Robots Will Change Us*. Cambridge, MA: MIT Press, 2002.

Brown, Rick, Bricknell Samantha. "What Is the Profile of Child Exploitation Material Offenders?" *Trends & Issues in Crime and Criminal Justice*, 564 (2018): 1-14.

Brown, Rick, Shelling Jane. "Exploring the Implications of Child Sex Dolls." *Trends & Issues in Crime and Criminal Justice*, 570 (2019): 1-13.

Bryson, Joanna J. "Robots Should Be Slaves." In *Close Engagements with Artificial Companions: Key Social, Psychological, Ethical and Design Issue*, edited by Yorick Wilks, 63-74. Amsterdam: John Benjamins, 2010.

Brynjolfsson, Erik, McAfee Andrew. *Race Against the Machine: How the Digital Revolution is Accellerating Innovation, Driving Productivity, and Irreversibly Transforming Employment and Economy*. Lexington (MA): Frontier Press, 2011.

Bullock, Caroline. "Attractive, Slavish and at Your Command: Is AI Sexist?" *BBC News*, December 2016. www.bbc.com/news/business-38207334.

Bullough, Vern L., McAnulty Richard D. "The Sex Trade: Exotic Dancing and Prostitution." In *Sex and Sexuality*, edited by Richard D. McAnulty, M. Michele Bumette, 299-320. Westport (Connecticut): Praeger, 2006.

Bumby, Kurt M. "Assessing the Cognitive Distortions of Child Molesters and Rapists: Development and Validation of the MOLEST and RAPE Scales." *Sexual Abuse: A Journal of Research and Treatment*, 8 (1996): 37-54.

Burgar, Charles G., Lum Peter S., Shor Peggy C., Van der Loos H.F. Machiel. "Development of Robots for Rehabilitation Therapy: The Palo Alto VA/Stanford Experience." *Journal of Rehabilitation Research and Development*, 37 (2000): 663-673.

Burnett, Dean. "Is the Internet Killing Our Brains?" *The Guardian*, October 8, 2016.

Burns, Lawrence D. "Sustainable Mobility: A Vision of Our Transport Future." *Nature*, 497 (2013): 181-182.

Burt, Martina R. "Cultural Myths and Supports for Rape." *Journal of Personality and Social Psychology*, 38 (1980): 217–230.

Cabibihan, John-John, Javed Hifza, Ang Marcelo H., Aljunied Sharifah Mariam. "Why Robots? A Survey on the Roles and Benefits of Social Robots for the Therapy of Children with Autism." *International Journal of Social Robotics*, 5 (2013): 593-618.

Calinon, Sylvain, Billard Aude. "Learning of Gestures by Imitation in a Humanoid Robot." In *Imitation and Social Learning in Robots, Humans and Animals*, edited by Chrystopher L. Nehaniv, Kerstin Dautenhahn, 153-177. Cambridge, UK: Cambridge University Press, 2007.

Carabellese, Felice, Candelli Chiara, La Tegola Donatella, Catanesi Roberto. "Fantasie sessuali, disturbi organici, violenze sessuali." *Rassegna Italiana di Criminologia*, 2 (2010): 347-360.

Carlstedt, Mathias, Bood Sven A., Norlander Torsten. "The Affective Personality and Its Relation to Sexual Fantasies in Regard to the Wilson Sex Fantasy Questionnaire." *Psychology*, 2 (2011), 792-796.

Carpenter, Julie. "Deus Sex Machina: Loving Robot Sex Workers and the Allure of an Insincere Kiss." In *Robot Sex. Social and Ethical Implications*, edited by John Danaher, Neil McArthur, 261-287. Cambridge: The MIT Press, 2017.

Carr, Nicholas. *The Shallows: How the Internet is Changing the Way We Think, Read and Remember.* New York: W.W. Norton & Company, 2011.

Casalini, Brunella. "Disabilità, immaginazione e cittadinanza sessuale." *Ethics & Politics*, 15, no. 2 (2013): 301-320.

Casetta, Elena. "Sesso." In *Manifesto per un nuovo femminismo*, edited by M. G. Turri, 167-181. Milano: Mimesis, 2013.

Casini, Stefano. "Non solo badanti e camerieri, i robot diventeranno anche compagni di 'piacere' e faremo sesso con loro." *Corriere della Sera Online*. March 1, 2018.

Chamayou, Gregoire. *A Theory of Drone*. New York: New Press, 2015.

Cheok, Adrian David, Karunanayaka Kasun, Zhang Emma Yann. "Lovotics." In *Robot Ethics 2.0. From Autonomous Cars to Artificial Intelligence*, edited by Patrick Lin, Keith Abney and Ryan Jenkins, 193-213. Oxford: Oxford University Press.

Cheok, Adrian David, Levy David, Karunanayaka Kasun. "Lovotics: Love and Sex with Robots". In *Emotion in Games*, edited by Kostas Karpouzis, Georgios N. Yannakakis, 303-328. Cham (Switzerland): Springer, 2016.

Cheok, Adrian David, Levy David, eds.. *Love and Sex with Robots. Third International Conference, LSR 2017.* Cham (Switzerland): Springer, 2018.

Coeckelbergh, Mark. "Personal Robots, Appearance, and Human Good: A Methological Reflection on Roboethics." *Int J Soc Robot*, 1 (2009): 217-221.

Coeckelbergh, Mark. "Artificial Companions: Empathy and Vulnerability Mirroring in Human-Robot Relations." *Studies in Ethics, Law and Technology* 4, no. 3 (2010): 1-17.

———. "Health Care, Capabilities, and AI Assistive Technologies." *Ethical Theory and Moral Practice* 13, no. 2 (2010): 181-190.

———. "Robot Rights? Towards a Social-relational Justification of Moral Consideration." *Ethics Inf Technol* 12, no. 3 (2010), 209-221.

———. "Humans, Animals, and Robots: A Phenomenological Approach to Human-Robot Relations." *Int J. Soc. Robot*, 3 (2011): 197-204.

Corner, Natalie. "Sex Doll the Size of a Child Being Made in a Japanese Factory Reduces a TV Presenter to Tears – as the Manufacturer Admits Customers Decide the Robots' 'Age'." *DailyMail*, April 10, 2018.

Corporate Vehicle Observatory. *Le auto a guida autonoma: siamo già nel futuro?* Scandicci (Fi), 2016.

Costescu, Cristina A., David Daniel O. "Attitudes Toward Using Social Robots in Psychotherapy." *Transylvanian Journal of Psychology*, 15 (2014): 3-20.

Costescu, Cristina A., Vanderborght Bram, David Daniel O. "The Effects of Robot-Enhanced Psychotherapy: A Meta-Analysis." *Review of General Psychology*, 18 (2014): 127-136.

Craig, Leam A., Dixon Louise, Gannon Theresa A., eds. *What Works in Offender Rehabilitation. An Evidence-Based Approach to Assessment and Treatment*. Chichester, UK: Wiley-Blackwell, 2013.

Couric, Katie. "You Can Soon Buy a Sex Robot Equipped with Artificial Intelligence for about $20.000." *ABC News*, April 25, 2018. https://www.youtube.com/watch?v=-cN8sJz50Ng.

Cox-George, Chantal, Bewley Susan, "I, Sex Robot: the Health Implications of the Sex Robot Industry." *BMJ Sexual & Reproductive Health*, 44 (2018):161-164.

Crawford, Neta C. *Accountability for Killing: Moral Responsibility in America's Post 9/11 Wars* (219-385). Oxford: Oxford University Press, 2014.

Danaher, John. "Sex Work, Technological Unemployment and the Basic Income Guarantee." *Journal of Evolution and Technology* 24, no.1 (2014): 113-130.

———. "Should Be Thinking about Robot Sex?" In *Robot Sex. Social and Ethical Implications*, edited by John Danaher and Neil McArthur, 3-14. Cambridge (Massachusetts): The MIT Press, 2017a.

—. "The Symbolic-Consequences Argument in the Sex Robot Debate." In *Robot Sex. Social and Ethical Implications*, edited by John Danaher, Neil McArthur, 103-131, Cambridge (Massachusetts): The MIT Press, 2017b.

—. "Robotic Rape and Robotic Child Sexual Abuse: Should They be Criminalised?" *Criminal Law and Philosophy* 11 (2017c): 71-95.

—. "Embracing the Robot." *Aeon*, March 19, 2018. https://aeon.co/essays/programmed-to-love-is-a-human-robot-relationship-wrong.

—. "Why We Should Create Artificial Offspring: Meaning and The Collective Afterlife." *Science and Engineering Ethics*, 24, 4 (2018): 1097-1118.

—. "The Philosophical Case for Robot Friendship." *Journal of Posthuman Studies* 3, no. 1 (2019): 5-24.

—. "Welcoming Robots into the Moral Circle: A Defence of Ethical Behaviourism." *Science and Engineering Ethics* (2019):1-17.

Danaher, John, Earp Brian, Sandberg Anders. "Should We Campaign Against Sex Robots?" In *Robot Sex. Social and Ethical Implications*, edited by John Danaher, Neil McArthur, 47-71. Cambridge (Massachusetts): The MIT Press, 2017.

Danaher, John, McArthur Neil, eds. *Robot Sex. Social and Ethical Implications*. Cambridge (Massachusetts): The MIT Press, 2017.

Darling, Kate. "Extending Legal Protection to Social Robots. The Effect of Anthropomorphism, Empathy and Violent Behavior Towards Robotic Objects." In *Robot Law*, edited by Ryan A. Calo, Michael Froomkin, Ian Kerr, 213-231. Cheltenham: Edward Elgar Publishing, 2016.

David, Daniel O., Matu Silviu A., David Oana Alexandra. "Robot-Based Psychotherapy: Concepts Development, State of the Art, and New Directions." *International Journal of Cognitive Therapy* 7 (2014): 192-210.

De Boer, Tracy. "Disability and Sexual Inclusion." *Hypatia* 30, no. 1 (2015): 66-81.

Dekkers, Midas. *Dearest Pet: On Bestiality*. London: Verso Books, 1994.

Delaunay, Nicolas. "Call for 'Virtual' Child Pornography to Be Legalised." *The Sydney Morning Herald*, November 20, 2012.

Delvaux, Mady. *Draft Report with Recommendations to the Commission on Civil Law Rules on Robotics*, 2017.

Demiris, Yiannis, Matthew Johnson. "Simulation Theory of Understanding Others: a Robotics Perspective." In *Imitation and Social Learning in Robots, Humans and Animals*, edited by Chrystopher L. Nehaniv, Kerstin Dautenhahn, 89-102. Cambridge, UK: Cambridge University Press, 2007.

Dennis, Elissa, Rouleau, Joanne-Lucine, Renaud Patrice, Nolet Kevin, Saumur Chantal. "A Pilot Development of Virtual Stimuli Depicting Affective Dispositions for Penile Plethysmography Assessment of Sex Offenders." *Canadian Journal of Human Sexuality* 23 (2014): 200-208.

De Sade. *Justine, or the Misfortunes of Virtue*. Translated by John Phillips, Oxford: Oxford University Press, 2012.

——. *Juliette*. Translated by Austryn Wainhouse, New York: Grove Press, 1968.

Devlin, Amanda, Lake Emma. "Robot Romps. What Is a Robot Sex Doll, What Is the Sex Robot Brothel in Paris and How Much Do They Cost?" *The Sun online*, March 7, 2018.

Devlin, Kate. *Turned on. Science, Sex and Robots*. London: Bloomsbury, 2018.

Diamond, Milton. "Pornography, Public Acceptance and Sex Related Crime: A Review." *Int J Law Psychiatry* 32, no. 5 (2009): 304-314.

Diamond, Milton, Uchiyama Ayako. "Pornography, Rape, and Sex Crimes in Japan." *International Journal of Law and Psychiatry* (1999): 1-22.

Diamond, Milton, Jozifkova Eva, Weiss Petr. "Pornography and sex crimes in the Czech Republic." *Arch Sex Bahev* 40, no. 5 (2011): 1037-1043.

Di Nucci, Ezio. "Sex Robots and the Rights of the Disabled." In *Robot Sex. Social and Ethical Implications*, edited by John Danaher, Neil McArthur, 73-88. Cambridge (Massachusetts): The MIT Press, 2017.

Döring, Nicola, Pöschl Sandra. "Sex Toys, Sex Dolls, Sex Robots: Our Under-Researched Bed-Fellows." *Sexologies* 27, no. 3 (2018): e51-e55.

Driver, Julia. *Uneasy Virtue*. Cambridge: Cambridge University Press, 2001.

Duggan, Lisa, Hunter Nan and Vance Carole, "False Promises: Feminist Antipornography Legislation.", *NYLS Law Review*, 38, 1, 1993, pp. 133-163.

Dumouchel, Paul, Damiano Luisa. *Living with Robots*. Harvard: Harvard University Press, 2017.

Dworkin, Andrea. *Pornography: Men Possessing Women*. London: Women's Press, 1981.

Dworkin, Andrea, MacKinnon Catherine Alice. *The Reasons Why: Essays on the New Civil Rights Law Recognizing Pornography as Sex Discrimination*. New York: Women Against Pornography, 1985.

Earp, Brian D. "Prostitution, Harm, and Disability: Should Only People with Disabilities Be Allowed to Pay for Sex?" *Journal of Medical Ethics*, 41, no. 6 (2015), e-letter.

——. "Prostitution and Disability." In *Philosophers Take on the World*, edited by David Edmonds, 99-103. Oxford: Oxford University Press, 2016.

Earp, Brian D., Moen Ole Martin. "Paying for Sex – Only for People with Disabilities?" *Journal of Medical Ethics* 42, no. 1 (2016): 54-56.

Earp, Brian D., Sandeberg Anders, Savulescu Julian. "The Medicalization of Love." *Cambridge Quarterly of Healthcare Ethics* 24 (2015): 323-336.

Earp, Brian D., Sandeberg Anders, Savulescu Julian. "The Medicalization of Love. Response to Critics." *Cambridge Quarterly of Healthcare Ethics*, 25 (2016): 759-771.

Earp, Brian D., Savulescu Julian. "Love Drugs: Why Scientistis Should Study the Effects of Pharmaceuticals on Human Romantic Relationships." *Technology in Society* 52, no. 1 (2018):10-16.

Earp, Brian D., Wudarczyk Olga A., Sandberg Anders, Savulescu Julian. "If I Could Just Stop Loving You: Anti-Love Biotechnology and the Ethics of a Chemical Breakup." *Am J Bioeth* 13, no. 11 (2013): 3-17.

Earp, Brian D., Wudarczyk Olga A., Savulescu Julian. "Addicted to Love: What Is Love Addiction and When Should It Be Treated?" *Philosophy, Psychiatry, & Psychology* 24, no. 1 (2017): 77-92.

Easton, Dossie, Hardy Janet W.. *Ethical Slut: A Practical Guide to Polyamory, Open Relationships & Other Adventures*. Emeryville (CA): Greenery Press, 1997.

Edwards, Lilian, Waelde Charlotte. *Law and Internet. Regulating Cyberspace*. Oxford: Hart Publishing, 1997.

Eke, Angela W., Seto Michael C., Williams Jannette. "Examining the Criminal History and Future Offending of Child Pornography Offenders: An Extended Prospective Follow-Up Study." *Law and Human Behavior* 35 (2011): 466-478.

Elliott, Larry. "Robots Will Take Our Jobs. We'Better Plan Now, Before It Is Too Late". *The Guardian*, February 1, 2018.

Eskens, Romy. "Is Sex with Robots Rape?" *Journal of Practical Ethics* 5, no. 2 (2017): 62-76.
European Parliament Resolution. *Civil Law Rules on Robotics*. European Parliament resolution of 16 February 2017 with recommendations to the Commission on Civil Law Rules on Robotics (2015/2103(INL)). P8_TA (2017) 0051, 2017.
Facchin, Federica, Giussy Barbara, Cigoli Vittorio. "Sex Robots: the Irreplaceable Value of Humanity." *BMJ* (2017): 358: j3790.
Fahs, Breanne, Swank Erik. "Adventures with the 'Plastic Man': Sex Toys, Compulsory Heterosexuality, and the Politics of Women's Sexual Pleasure." *Sexuality & Culture* 17 (2013): 666-685.
Farley, Melissa, Howard Barkan. "Prostitution, Violence against Women, and Posttraumatic Stress Disorder." *Women Health*, 27 (1998): 37-49.
Faustino, Maria João. "Rebooting an Old Script by New Means: Teledildonics – The Technological Return to the 'Coital Imperative'." *Sexuality & Culture*, 22, no. 1 (September 21, 2017): pp. 243-257.
Feelgood, Steven, Cortoni Franca, Thompson Anthony. "Sexual Coping, General Coping and Cognitive Distortions in Incarcerated Rapists and Child Molesters." *Journal of Sexual Aggression* 11 (2005): 157-170.
Fellous, Jean-Marc, Arbib Michael A. *Who Needs Emotions? The Brain Meets the Robot*. Oxford: Oxford University Press, 2005.
Ferguson, Anthony. *The Sex Doll. A History*. North Caroline: McFarland & Company, Inc., 2010.
Ferguson, Christopher J. "Evidence for Publication Bias in Video Game Violence Effects Literature: A Meta-Analytic Review." *Aggression and Violent Behavior* 12 (2007a): 470-482.
———. "The Good, the Bad and the Ugly: A Metaanalytic Review of Positive and Negative Effects of Violent Video Games." *Psychiatric Quarterly* 78, no. 4 (2007b): 309-316.
Ferguson, Christopher J., Beaver Kevin M. "Who's Afraid of the Big, Bad Video Game? Media-Based Moral Panics." In *Psychology of Fear, Crime and the Media*, edited by Derek Chadee, 240-252. London: Routledge, 2015.
Ferguson, Christopher J., Hartley Richard D. "The Pleasure is Momentary…the Expense Damnable? The Influence of Pornography on Rape and Sexual Assault." *Aggression and Violent Behavior* 14 (2009): 323–329.

Ferguson, Christopher J., Kilburn John. "The Public Health Risks of Media Violence: A Meta Analytic Review." *J. Pediatr* 154, no. 5 (2009): 759-763.

———. "Much Ado About Nothing: The Misestimation and Overinterpretation of Violent Video Game Effects in Eastern and Western Nations: Comment on Anderson et al." *in* Psychological Bulletin 136, no. 2 (2010):174-8.

Forbes. "How the 'Niche' Sex Toy Market Grew into an Unstoppable $15B Industry." 2016. Retrieved from http://www.forbes.com/sites/janetwburns/2016/07/15/adult-expo-founders-talk-15b-sext oy-industry-after-20-years-in-the-fray/$508e13938a19/.

Ford, Martin. *Rise of the Robot: Technology and the Threat of a Jobless Future.* New York: Basic Books, 2015.

Fornari, Ugo. *Trattato di psichiatria forense* (6ª ed.). UTET: Torino, 2015.

———. *Trattato di psichiatria forense* (7ª ed.; I e II tomo). UTET: Torino, 2018.

Fosch-Villaronga Eduard, Poulsen Adam. "Exploring the Potential Use of Sexual Robot Technologies for Disabled and Elder Care." forthcoming.

Frank, Lily E., Nyholm Sven R. "Robot Sex and Consent." *Artificial Intelligence and Law* 25 (2017): 305-323.

Freilone, Franco, Dotta Marianna, Veggi Sara, Zara Georgia. "La psicologia delle relazioni sintetiche. Affettività e sessualità 'alternative' con i sexbot." *NEU*, 3 (2018): 55–67.

Frischmann, Brett, Selinge Evan. *Re-engineering humanity*. Cambridge, UK: Cambridge University Press, 2018.

Garofalo Geymonat, Giulia. "Lavoro sessuale in Europa." *Il Mulino*, 2 (2011): 291-299.

———. *Vendere e comprare sesso*. Bologna: Il Mulino, 2014.

Garofalo Geymonat, Giulia. "Oltre il dibattito pubblico, ma non oltre la critica: pratiche di assistenza sessuale in Europa." In *Loveability. L'assistenza sessuale per le persone con disabilità*, edited by M. Ulivieri, 99-114. Roma: Erickson, 2014.

Gee, Dion G., Ward Tony, Eccleston Lynne. "The Function of Sexual Fantasies for Sexual Offenders: A Preliminary Model." *Behavior Change* 20 (2003): 44-60.

Gee, Dion G., Devilly Grant J., Ward Tony. "The Content of Sexual Fantasies for Sexual Offenders." *Sexual Abuse: A Journal of Research and Treatment* 16 (2004): 315-331.

Gelderblom, Gert Jan, Bemelmans Roger, Spierts Nadine, Jonker Pieter, De Witte Luc. "Development of PARO Interventions for Dementia Patients in Dutch Psycho-Geriatric Care." In *Social Robotics. ICSR: International Conference on Social Robotics*, 253-258. Berlin Heidelberg: Springer, 2010.

Gentile, Douglas A. "Catharsis and Media Violence: A Conceptual Analysis." *Societies* 3 (2013): 491-510.

George, Robert P., Tollefsen Christopher. *A Defese of Human Life*. New York: Doubleday, 2008.

Goldman, Alan H. "Plain Sex." *Philosophy and Public Affairs* 6, no. 3 (1977): 267-287.

Gomes, Leonardo M., Wu Rita. "Neurodildo: A Mind-Controlled Sex Toy with E-stim Feedback for People with Disabilities." In *Love and Sex with Robots. Third International Conference, LSR 2017*, edited by Adrian David Cheok, David Levy, 65-82. Cham (Switzerland): Springer, 2018.

Goodal, Noah Joseph. "Machine Ethics and Automated Vehicles." In *Road Vehicle Automation*, edited by Gereon Meyer, Sven Beiker, 93-102. Springer Intenational Publishing, 2014a.

———. "Ethical Decision Making During Automated Vehicle Crashes." *Transportation Research Record: Journal of the Transportation Research Board*, No. 2424. Transportation Board of the National Academies, Washington D.C., 58-65, 2014b.

Greenwald, Anthony G., Banaji Mahzarin R., Rudman Laurie A., Farnham Shelly D., Nosek Brian A., Mellott Deborah S. "A Unified Theory of Implicit Attitudes, Stereotypes, Self-esteem, and Self-Concept." *Psychological Review* 109 (2002) : 3–25.

Grundmann, Dorit, Krupp Jurian, Scherner Gerold, Amelung Till, Beier Klaus. "Stability of Self-reported Arousal to Sexual Fantasies Involving Children in a Clinical Sample of Pedophiles and Hebephiles." *Archives of Sexual Behavior*, 45 (2016): 1153-1162.

Gunkel, David J. "A Vindication of the Rights of Machines." *Philos Technol* 27, no. 1 (2014): 113–132.

———. *Robot Rights*. Cambridge (Massachusetts): The MIT Press, 2018.

Gutiu, Sinziana. "Sex Robots and Roboticization of Consent." *We Robot Conference. Miami*, 1-24. 2012.

———. "The Robotization of Consent." In *Robot Law*, edited by Ryan Calo, A, Michael Froomkin, Ian Kerr, 186-212. Cheltenham: Edward Elgar Publishing, 2016.

Hanson, R. Karl, Bourgon Guy, Helmus Leslie, Hodgson Shannon. *A Meta-Analysis of the Effectiveness of Treatment for Sex Offenders: Risk, Need, and Responsivity*. Ottawa, ON: Public Safety Canada, 2009.

Hanson, R. Karl, Harris Andrew J., Scott Terri-Lynne, Helmus Leslie. *Assessing the Risk of Sexual Offenders on Community Supervision: The Dynamic Supervision Project*. Public Safety Canada, 2007.

Hanson, R. Karl, Harris Andrew J. R., Helmus Leslie, Thornton David. "High-Risk Sex Offenders May Not Be High Risk Forever." *Journal of Interpersonal Violence* 29 (2014): 2792-2813.

Hanson, R. Karl, Thornton David. *Static-99: Improving Actuarial Risk Assessments for Sex Offenders*. User Report 99-02. Ottawa: Department of the Solicitor General of Canada, 1999.

———. "Improving Risk Assessments for Sex Offenders: A Comparison of Three Actuarial Scales." *Law and Human Behavior* 24 (2000): 119-136.

Harari, Yuval Noah. "The Meaning of Life in a World without Work." *The Guardian*, May 8, 2017.

Hauskeller, Michael. *Sex and the Posthuman Condition*. London: Palgrave, 2014.

———. "Automatic Sweethearts for Transhumanists." In *Robot Sex. Social and Ethical Implications*, edited by John Danaher, Neil McArthur, 203-218. Cambridge (Massachusetts): The MIT Press, 2017.

Heinrich, Liesi M., Gullone Eleonora. "The Clinical Significance of Loneliness. A literature Review." *Clinical Psychology Review* 26, no. 6 (2006): 695-718.

Herbenick, Debby, Sanders Stephanie, Reece Michael, Doodge Brian. "Prevalence and Characteristics of Vibrator Use by Women in the United States: Results from a Nationally Representative Study". *J Sex Med.* 8, no. 7 (2009): 1857-66.

———. "Women's Vibrator Use in Sexual Partnerships: Results from a Nationally Representative Survey in the United States." *Journal of Sex & Marital Therapy* 36, no. 1 (2010): 49-65.

Herzfeld, Noreen. "Religious Perspectives on Sex with Robots." In *Robot Sex. Social and Ethical Implications*, edited by John Danaher, Neil McArthur, 91-101. Cambridge (Massachusetts): The MIT Press, 2017.

Hevelke, Alexander, Nida-Rumelin Julian. "Responsibility for Crashes of Autonomous Vehicles: An Ethical Analysis." *Sci Eng Ethics* 21 (2015): 619-630.

Howitt, Dennis. "What Is the Role of Fantasy in Sex Offending?" *Criminal Behaviour and Mental Health* 14 (2004): 182-188.

Hudson, Sthephen M., Wales David S., Bakker Leon, Ward Tony. "Dynamic Risk Factors: The Kia Marama Evaluation." *Sexual Abuse: A Journal of Research and Treatment* 14 (2002): 101-117.

Huesmann, L. Rowell. "Nailing the Coffin Shut on Doubts that Violent Video Games Stimulate Aggression: Comment on Anderson et al (2010)". *Psychological Bulletin*, 136 (2010): 179–181.

Huffingtonpost, *Robot Sex Poll Reveals Americans' Attitudes About Robotic Lovers, Servants, Soldiers*, October 4, 2013.

Hume, David. *A Treatise of Human Nature (1739-1741)*, edited by L.A. Selby-Bigge, P.H. Nidditch, Oxford: Clarendon Press, 1960.

Jones, Jennifer C., Barlow David H. "Self-Reported Frequency of Sexual Urges: Fantasies and Masturbatory Fantasies in Heterosexual Males and Females." *Archives of Sexual Behavior* 19 (1990): 269-279.

Jung, Sandy. *RNR Principles in Practice in the Management and Treatment of Sexual Abusers*. Brandon, Vermont: Safer Society, 2017.

Kachouie, Reza, Sedighadeli Sima, Khosla Rajiv, Chu Mei-Tai. "Socially Assistive Robots in Elderly Care: A Mixed-Method Systematic Literature Review." *International Journal of Human-Computer Interaction* 30 (2014): 369-393.

Kanamori, Masao, Suzuki Misao, Tanaka Mizue. "Maintenance and Improvement of Quality of Life Among Elderly Patients Using a Pet-Type Robot." *Japanese Journal of Geriatrics* 39 (2002): 214-218.

Kingston, Drew A., Fedoroff Paul, Firestone Philip, Curry Susan, Bradford John M. "Pornography Use and Sexual Aggression: The Impact of Frequency and Type of Pornography Use on Recidivism Among Sexual Offenders." *Aggressive Behavior* 34 (2008): 341-351.

Kipnis Laura. *Bound and Gagged: Pornography and Politics of Fantasy in America*. New York: Grove Press, 1996.

Kirkendall, Lester A., McBride Leslie G. "Preadolescent and Adolescent Imagery and Sexual Fantasies: Beliefs and Experiences". In *Handbook of Sexology: Childhood and Adolescent Sexology*, edited by Mary Elizabeth Perry, vol. 7, 263-287. Amsterdam: Elsevier, 1990.

Knafo, Danielle, Yoram Jaffe. "Sexual Fantasizing in Males and Females." *Journal of Research in Personality* 18 (1984): 451-462.

Kolivand, Hoshang, Rad Abdoulvahab Ehsani, Tully David. "Virtual Sex: Good, Bad or Ugly?" In *Love and Sex with Robots. Third International Conference, LSR 2017*, edited by Adrian David Cheok, David Levy, 26-36. Cham (Switzerland): Springer, 2018.

Kontula, Osmo. *Between Sexual Desire and Reality: The Evolution of Sex in Finland the Family Federation of Finland.* Helsinki: The Population Research Institute (D49), 2009.

Krugman, Paul. "Robots and Robber Barons." *New York Times*, December 9, 2012.

——. "Sympathy for the Luddites." *New York Times*, June 13, 2013.

Kühn, Simone, Kugler Dimitrij, Schmalen Katharina, Weichenberger Markus. "The Myth of Blunted Gamers: No Evidence for Desensitization in Empathy for Pain after a Violent Video Game Intervention in a Longitudinal fMRI Study on Non-Gamers." *Neuro-Signas*, February 6, 2018a.

Kühn, Simone, Kugler Dimitrij Tycho, Schmalen Katharina, Weichenberger Markus, Witt Charlotte, Gallinat Jürgen. "Does Playing Violent Video Games Cause Aggression? A Longitudinal Intervention Study." *Molecular Psychiatry*, March 13, 2018b.

LaGrandeur, Kevin, Hughes James J.. *Surviving the machine Age: Intelligent Technology and the Transformation of Human Work.* Cham: Palgrave Macmillan, 2017.

Langewiesche, William. *Esecuzioni a distanza.* Milano: Adelphi, 2011.

Langton, Rae. "Speech Acts and Unspeakable Acts." *Philosophy & Public Affairs* 22, no. 4 (1993): 292-330.

Langton, Rae, West Caroline. "Scorekeeping in a Pornographic Language Game." *Australasian Journal of Philosophy* 77, no. 3 (1999): 303-319.

Laue, Cheyenne. "Familiar and Strange: Gender, Sex, and Love in the Uncanny Valley." *Multimodal Technologies and Interact* 1, no. 2 (2017): 1-11.

Lecaldano, Eugenio. *Prima lezione di filosofia morale.* Roma: Laterza, 2010.

Lee, Jason. *Sexbots. The Future of Desire.* Switzerland: Palgrave MacMillan, 2017.

Leitenberg, Harold, Henning Kris. "Sexual Fantasy." *Psychological Bulletin*, 117 (1995): 469-496.

Levy, David. *Emotional Relationship with Robotic Companions.* Genova: EURON Workshop on Roboethics, 2006.

———. "Robot Prostitutes as Alternative to Human Sex Workers." *Proceedings of the IEE-RAS International Conference an Robotics and Automation (ICRA 2007)*, April 10-14 2007 Roma.

———. *Love and Sex with Robots: The Evolution of Human-Robot Relationships* (2007). New York, USA: Harper, 2008.

———. "The Ethics of Robot Prostitutes." In *Robot Ethics. The Ethical and Social Implications of Robotics*, edited by Patrick Lin, Keith Abney, George A. Bekey, 223-231. Cambridge, Massachusetts: The MIT Press, 2012.

———. "Roxxxy the "Sex Robot"—Real or Fake?" *Lovotics* 1 (2013): 1-4.

———. "Why Not Marry a Robot?" In *Love and Sex with Robots*, edited by Adrian David Cheok, Kate Devlin, D. Levy, 3-13. Second International Conference, LSR 2016 London, UK, December 19-20, 2016 Revised Selected Papers. London: Springer International Publishing, 2017.

Libin, Alexander V., Libin Elena V. "Person–Robot Interactions from the Robopsychologists' Point of View: The Robotic Psychology and Robotherapy Approach." *Proceedings of the IEEE* 92 (2004): 1789–1803.

Liberati, Nicola. "Being Riajuu: A Phenomenological Analysis of Sentimental Relationships with 'Digital Others'." In *Love and Sex with Robots. Third International Conference, LSR 2017*, edited by Adrian David Cheok, David Levy, 12-25. Cham (Switzerland): Springer, 2018.

Lin Patrick, Abney Keith, Bekey George A., eds. *Robot Ethics: The Ethical and Social Implications of Robotics*. Cambridge: MIT Press, 2011.

Lin, Patrick. "Why Ethics Matters for Autonomous Cars." In *Autonomous Driving. Technical, Legal and Social Aspects*, edited by Markus Maurer, J. Christian Gerdes, Barbara Lenz, Hermann Winner, 69-85. Berlin: Springer 2016.

Looman, Jan. "Sexual Fantasies of Child Molesters." *Canadian Journal of Behavioral Sciences* 27 (1995): 321-332.

Lösel, Friedrich, Schmucker Martin. "The Effectiveness of Treatment for Sexual Offenders: A Comprehensive Meta-Analysis." *Journal of Experimental Criminology* 1 (2005): 117-146.

Lösel, Friedrich, Schmucker Martin. "Treatment of Sex Offenders." In *Encyclopedia of Criminology and Criminal Justice*, edited by G. Bruinsma & D. Weisburd, 5323-5332. New York: Springer, 2014.

Luck, Morgan. "The Gamer's Dilemma: An Analysis of the Arguments for the Moral Distinction between Virtual Murder and Virtual Pedophilia." *Ethics and Information Technology* 11 (2009): 31-36.

Lunceford, Brett. "The Body and the Sacred in the Digital Age: Thoughts on Posthuman Sexuality." *Theology and Sexuality* 15, no. 1 (2009): 77-96.

———. "Telepresence and the Ethics of Digital Cheating." *Explorations in Media Ecology*, 12, no. 1-2 (2013): 7-26.

MacKellar, Jean Scott. *Rape: The Bait and the Trap. A Balanced, Humane, Up-to-Date Analysis of Its Causes and Control.* New York: Crown Publishers, 1975.

Magnani, Alberto. "In Italia la tassa sui robot non funzionerebbe: ecco perché." *Il Sole 24 ore*, February 8, 2018.

Mann, Ruth E., Beech Anthony R. "Cognitive Distortions, Schemas, and Implicit Theories." In *Sexual Deviance: Issues and Controversies*, edited by Tom Ward, D. Richard Laws, & Stephen M. Hudson, 135-153. Thousand Oaks, CA: Sage, 2003.

Mann, Ruth E., Hanson R. Karl, Thornton David. "Assessing Risk for Sexual Recidivism: Some Proposals on the Nature of Psychologically Meaningful Risk Factors." *Sexual Abuse: A journal of Research and Treatment* 22 (2010): 191-217.

Mann, Ruth E., Shingler Jo. "Schema-Driven Cognition in Sexual Offenders: Theory, Assessment and Treatment." In *Sexual offender treatment: Controversial Issues*, edited by William Lamont Marshall, Yolanda M. Fernandez, Liam E. Marshall & Geris A. Serran, 173-185. Chichester, UK: John Wiley & Sons, Ltd, 2006.

Mann, Ruth E., Webster Stephen D., Wakeling Helen C., Keylock Helen. "Why Do Sexual Offenders Refuse Treatment?" *Journal of Sexual Aggression* 19 (2013): 191-206.

Maras, Marie-Helen, Shapiro Lauren R. "Child Sex Dolls and Robots: More than Just an Uncanny Valley." *Journal of Internet Law* (2017): 3-21.

Markey, Patrick M., Markey Charlotte N., French Juliana E. (2015). Violent Video Games and Real-World Violence: Rhetoric Versus Data. *Psychology of Popular Media Culture*, 4(4), 277-295.

Marrone, Pierpaolo. "Apocalissi sessuali: gli incubi di una visionaria illustrati con i sogni della robotica." In *Pop-Sophia. 12 Ingressi (senza omaggi) alla filosofia*, 57-70. Milano: Mimesis, 2018.

Marshall, William Lamont. "Intimacy, Loneliness and Sexual Offenders." *Behavioural Research and Therapy* 27 (1989): 491-503.

Marshall, William Lamont, Marshall Liam E., Sachdev Sarah, Kruger Raina-Lianne. "Distorted Attitudes and Perceptions, and their

Relationship with Self-Esteem and Coping in Child Molesters." *Sexual Abuse: A Journal of Research and Treatment* 15 (2003): 171–181.

Marshall, William Lamont, Marshall Liam E., Serran Gerris A., O'Brien Matt D., eds. *Rehabilitating Sexual Offenders: A Strength-Based Approach.* Washington, DC: American Psychological Association, 2011.

McArthur, Neil. "The Case for Sexbots." In *Robot Sex. Social and Ethical Implications*, edited by John Danaher, Neil McArthur, 31-45. Cambridge (Massachusetts): The MIT Press, 2017.

McCarthy, Jennifer A. "Internet Sexual Activity: A Comparison Between Contact and Non-Contact Child Pornography Offenders." *Journal of Sexual Aggression* 16, no. 2 (2010): 181-195.

McCormick, Matt, "Is It Wrong to Play Violent Video Games?" *Ethics and Information Technology* 3 (2001): 277-287.

McDorman, Karl F. "Androids as an Experimental Apparatus: Why Is There an Uncanny Valley and Can We Exploit It?" In *CogSci-2005 Workshop: Toward Social Mechanisms of Android Science* (108-118), Stresa, July 25-26, 2005.

McElroy, Wendy. "A Feminist Defense of Pornography." *Free Inquiry Magazine* 17, no. 4 (1997).

McEwan, Ian. *Machines Like Me*. London, UK: Jonathan Cape, 2019.

MacKinnon, Catherine. *Only Words*. London: Harper Collins, 1995.

Meston, Cindy M., Buss David M. "Why Human Have Sex." *Arch Sex Behav* 36 (2007): 477-507.

Mews, Aidan, Di Bella Laura, Purver Mark. *Impact evaluation of the prison-based Core Sex Offender Treatment Programme*. London: Ministry of Justice Ministry of Justice Analytical Series, 2017.

Micklethwaite, Jamie. "Porn Star Behind World's Most Expensive Sex Robot." *Daily Star*, August 1, 2017.

Midgley, Mary. "Brutality and Sentimentality." *Philosophy* 54, n. 209 (1979): 385-389.

Migotti, Mark, Wyatt Nicole. "On the Very Idea of Sex with Robots." In *Robot Sex. Social and Ethical Implications*, edited by John Danaher, Neil McArthur, 15-28. Cambridge (Massachusetts): The MIT Press, 2017.

Miley, Jessica. "Sex Robot Samantha Gets an Update to Say 'No' If She Feels Disrespected or Bored." *Interesting Engineering*, June 29, 2018.

Moen, Ole Martin. "Is Prostitution Harmful?" *Journal Of Medical Ethics* (2012): 1-9.

——. "Prostitution and Sexual Ethics: A Reply to Westin." *Journal of Medical Ethics* 40, no. 2 (2014): 88.

——. "Prostitution and Harm: A Reply to Anderson and McDougall." *Journal of Medical Ethics* 40, no. 2 (2014): 84-85.

Moen, Ole Martin, Sterri A.B. "Pedophilia and Computer-Generated Child Pornography." In *The Palgrave Handbook of Philosophy and Public Policy*, edited by D. Boonin, 369-381. Cham: Palgrave Macmillan, 2018.

Monto, Martin A. "Prostitution and Fellatio." *Journal of Sex Research* 58, no. 2 (2001): 140-145.

Moran, Rachel. *Paid For. My Journey through Prostitution*. New York & London: W. W. Norton & Company, 2015. epub

Moravec, Hans. *Robot: Mere Machine to Transcendent Mind.* Cambridge, MA: MIT Press, 2001.

Morgan, Seiriol. "Sex in the Head." *Journal of Applied Philosophy* 20, no. 1 (2003a): 1-16.

——. "Dark Desires." *Ethical Theory and Moral Practice* 6, no. 4 (2003b): 377-410.

Morgan, William. "Is It Possible to Give a Philosophical Definition of Sexual Desire?" *PhilonoUS*, 1 (2016): 47-58.

Mori, Masahiro. "The Uncanny Valley (1970)." *Energy* 7, no. 4 (2012): 33-35.

Moyle, Wendy, Kellett Ursula, Ballantyne Alison, Gracia Natalie. "Dementia and Loneliness: An Australian Perspective." *Journal of Clinical Nursing* 20, no. 9-10 (2011): 1445-1453.

Murphy, William D. "Assessment and Modification of Cognitive Distortions in Sex Offenders." In *Handbook of Sexual Assault: Issues, Theories, and Treatment of the Offender*, edited by W. L. Marshall, D. R. Laws, & Barbaree, pp. 331–342. NY: Plenum, 1990.

Neely, Erika L., "Machines and the Moral Community." *Philosophy and Technology* 27, no. 1 (2014): 97-111.

Nehaniv, Chrystopher L., Dautenhahn Kerstin, eds. *Imitation and Social Learning in Robots, Humans and Animals*. Cambridge, UK: Cambridge University Press, 2007.

Neidigh, Larry, Krop Harry. Cognitive Distortions Among Child Sexual Offenders. *Journal of Sex Education and Therapy* 18 (1992) : 208–215.

Nicolaci da Costa, Pedro. "I robot iniziano a fare I lavori dei 'colletti bianchi', ma non vi ruberanno il posto. Per ora." *Business Insider Italia*, April 6, 2017.

Niveau, Gérard. "Cyber-pedocriminality: Characteristics of a Sample of Internet Child Pornography Offenders." *Child Abuse & Neglect* 34 (2010): 570-575.

Nunes, Kevin L., Pettersen Catherine. "Competitive Disadvantage Makes Attitudes Towards Rape Less Negative." *Evolutionary Psychology* 9 (2011): 509–521.

Nunes, Kevin L., Pettersen Catherine, Hermann Chantal A., Looman Jan, Spape Jessica. "Does Change on the MOLEST and RAPE Scales Predict Sexual Recidivism?" *Sexual Abuse: A Journal of Research and Treatment*, 28 (2016): 427 –447.

Nussbaum, Martha. "'Whether from Reason or Prejudice': Taking Money for Sexual Service." *Journal of Legal Studies*, 27 (1998): 693-723.

Nyholm, Sven, Frank Lily E. "From Sex Robots to Love Robots: Is Mutual Love with a Robot Possible?" In *Robot Sex. Social and Ethical Implications*, John Danaher, Neil McArthur, 219-243. Cambridge (Massachusetts): The MIT Press, 2017.

Owsianik, Jenna, "State of Sex Robots: These Are the Companies Developing Robotic Lovers (Updated)." *Futureofsex*, November 16, 2017. https://futureofsex.net/robots/state-sex-robots-companies-developing-robotic-lovers/.

Palazzani, Laura. *Dalla bio-etica alla tecno-etica: nuove sfide al diritto*. Torino: Giappichelli, Torino, 2017.

Parfit, Derek. *Reason and Persons*. Oxford: Oxford University Press, 1984.

Patridge, Stephanie. "The Incorrigible Social Meaning of Video Game Imagery." *Ethics Inf Technol* 13 (2011): 303-312.

Paura, Roberto, Verso Francesco., eds. *Segnali dal futuro*. Napoli: Italian Institute for the Future, 2016.

Pax. *Deadly Decisions. 8 Objections to Killer Robots*. Utrecht, 2014.

Pervan, Susan, Hunter Mick "Cognitive Distortions and Social Self-Ssteem in Sexual Offenders." *Applied Psychology in Criminal Justice* 3 (2007): 75–91.

Petersen, Steve. "Is It Good for Them Too? Ethical Concern for the Sexbots?" In *Robot Sex. Social and Ethical Implications*, edited by John Danaher, Neil McArthur, 155-200. Cambridge (Massachusetts): The MIT Press, 2017.

Picard, Rosalin. *Affective Computing*. Cambridge, MA: MIT Press, 1997.
Piha, Samuel, Hurmerinta Leila, Sandberg Birgitta, Järvinen Elina. "From Filthy to healthy and Beyond: Finding the Boundaries of Taboo Destruction in Sex Toys Buying." *Journal of Marketing Management* (2018): 1078-1104.
Pinker, Steven. *The Better Angels of Our Nature: Why Violence Has Declined*. New York: Viking, 2011.
———. "Has the Decline of Violence Reversed since The Better Angels of Our Nature was Written?", 2017: 1-12. stevenpinker.com/files/pinker/files/has_the_decline_of_violence_reversed_since_the_better_angels_of_our_nature_was_written_2017.pdf
Piquero, Alex R., Farrington David P., Jennings Wesley G., Diamond Brie, Craig Jessica. "Sex Offenders and Sex Offending in the Cambridge Study in Delinquent Development: Prevalence, Frequency, Specialization, Recidivism and (Dis)Continuity over the Lifecourse." *Journal of Crime and Justice* 35 (2012): 412-426.
Plato. *The Symposium*. Cambridge: Cambridge University Press, 2008.
Poli, Patrizia, Morone Giovanni, Rosati Giulio, Masiero Stefano. "Robotic Technologies and Rehabilitation: New Tools for Stroke Patients' Therapy." *BioMed Research International* (2013): 1-8.
Porter, Roy, Lesley Hall. *The Facts of Life: the Creation of Sexual Knowledge in Britain, 1650-1950*. New Haven: Yale University Press, 1995.
Prigg, Mark. "Could Child Sex Robots Be Used to Treat Pedophiles? Researchers Say Sexbots Are Inevitable and Could Be Used 'Like Methadone for Drug Addicts'." *Dayly Mail Online*, July 16, 2014.
Primoratz, Igor. *Ethics and Sex*. London: Routledge, 1999.
Pugmire, David. *Sound Sentiments*. Oxford: Oxford University Press, 2005.
Quattrini, Fabrizio. "Assistenza sessuale: il progetto 'LoveGiver' per la formazione degli operatori." In *Loveability. L'assistenza sessuale per le persone con disabilità*, edited by Max Ulivieri, 61-91. Trento: Erickson, 2014.
Quattrini, Fabrizio, Ulivieri Max. "Cosa è (e cosa non è) l'assistenza sessuale." In *Loveability. L'assistenza sessuale per le persone con disabilità*, edited by Max Ulivieri, 51-59. Erickson, Trento, 2014.
Rachels, James. *Created by Animal. The Moral Implications of Darwinism*. Oxford: Oxford University Press, 1990.
Rambow, Jana, Elsner K., Feelgood Steven, Hoyer Jürgen. "Einstellungen zum kindesmissbrauch: Untersuchungen mit der

Bumby Child Molest Scale bei missbrauchs- und gewalttätern [Attitudes toward child abuse: Using the Bumby Child Molest Scale in studies of sexual and violent offenders.]." *Zeitschrift für Sexualforschung* 21 (2008): 341–355.

Renaud, Patrice, Chartier Sylvain, Rouleau Joanne-L., Proulx Jean, Décarie Jean, Trottier Dominique, Bradford John P., Fedoroff Paul, Bouchard Stephane. "Gaze Behavior Nonlinear Dynamics Index of Sexual Deviancy: Preliminary Results." *Journal of Virtual Reality and Broadcasting*, 6 (2009).

Ricca, Jacopo. "A Torino apre la prima casa di appuntamenti con "sex doll" per uomini e donne." *La Repubblica online*, August 8, 2018.

Richardson, Kathleen. "The 'Asymmetrcal' Relationship: Parallels between Prostitution and the Development of Sex Robots." *SIGCAS Computers & Society* 45, no. 3 (2015): 290-293.

———. "Sex Robot Matters." *IEEE Technology and Society Magazine*, June 2016: 46-53.

———. "Urgent – Why We Must Campaign to Ban Sex Dolls & Sex Robots." *Campaign against Sex Robots*, October 3, 2017.

Riegel, David L. "Effects on Boy-Attracted Pedosexual Males of Viewing Boy Erotica." *Archives of Sexual Behavior* 33, no. 4 (2004): 321-323.

Riek, Laurel D., Robinson Peter. "Using Robots to Help People Habituate to Visible Disabilities." *2011 IEEE International Conference on Rehabilitation Robotics Rehab Week Zurich*. ETH Zurich Science City, Switzerland, June 29 - July 1, 2011.

Rigotti, Carlotta, "Sex Robots: a Human Rights Discourse?" *OpenGlobalRights*, May 2019.

Robinson, Hayley, MacDonald Bruce, Kerse Ngaire, Broadbent Elizabeth. "The Psychosocial Effects of a Companion Robot: A Randomized Controlled Trial." *JAMDA* 30 (2013): 1-7.

Robinson, Hayley, MacDonald Bruce, Broadbent Elizabeth. "The Role of Healthcare Robots for Older People at Home: A Review." *International Journal of Social Robotics*, 6, no. 4 (2014): 575-591.

Rodogno, Raffaele. "Social Robots, Fiction, and Sentimentality." *Ethics and Information Technology* 18, no. 4 (2016): 257-268.

Rogers, Richard, Dickey Rob. "Denial and Minimization Among Sex Offenders: A Review of Competing Models of Deception." *Annals of Sex Research* 4 (1991) : 49–63.

Rousi, Rebekah. "Lying Cheating Robots – Robots and Infidelity." In *Love and Sex with Robots. Third International Conference, LSR 2017*, edited by Adrian David Cheok, David Levy, 51-64. Cham (Switzerland): Springer, 2018.

Royakkers Lamber, Van Est Rinie. *Just Ordinary Robots: Automation from Love to War*. Boca Raton: CRC Press, 2015.

Rutkin, Aviva. "Could Sex Robots and Virtual Reality Treat Paedophilia? Not Like Us." *NewScientist*, August 2, 2016.

Sanders, Teela. "Female Sex Workers as Health Educator with Men Who Buy Sex: Utilising Narratives of Rationalizations." *Social Sciences and Medicine* 62, no. 10 (2005): 2434-2444.

——. "The Politics of Sexual Citizenship: Commercial Sex and Disability." *Disability & Society* 22, no. 5 (2007): 439-455.

Sanders, Teela, Jane Pitcher, Maggie O'Neill. *Prostitution: Sex Work, Policy and Politics*. London: Sage, 2009.

Santoni de Sio, Filippo, Van Wynsberghe Aimee. "When Should We Use Care Robots? The Nature-of-Activities Approach." *Science and Engineering Ethics* 22, no. 6 (2016): 1745-1760.

Savulescu, Julian, Birks David. *Bioethics: Utilitarianism. Revised version of Julian Savulescu 2006*. Chichester, UK: eLS, John Wiley and Sons, 2012. http://wwwels.net..

Savulescu, Julian, Earp Brian D. "Neuroreductionism about Sex and Love." *Think (Lond)*, 13, no. 38 (2014): 7-12.

Scheutz, Matthias, Arnold Thomas. "Are We Ready for Sex Robots?" *Proceeding HRI 2016 The Eleventh ACM/IEEE International Conference on Human Robot Interaction*, 2016, 351-358.

Scheutz, Matthias, Arnold Thomas. "Intimacy, Bonding, and Sex Robots: Examining Empirical Results and Exploring Ethical Ramifications." In *Robot Sex. Social and Ethical Implications*, edited by John Danaher, Neil McArthur, 247-260. Cambridge (Massachusetts): The MIT Press, 2017.

Schick, Vanessa, Herbenick Debby, Rosenberger Joshia G., Reece Michael. "Prevalence and Characteristics of Vibrator Use among Women Who Have Sex with Women." *J Sex Med.*, 8, no. 12 (2011): 3306-15.

Schlesinger, Louis B.. *Serial Offenders. Current Thought, Recent Findings*. New York: CRC Press, 2000.

Schmucker, Martin, Lösel Friedrich. "Does Sexual Offender Treatment Work? A Systematic Review of Outcome Evaluations." *Psicothema* 20 (2008): 10-19.

Schmucker, Martin, Lösel Friedrich. "The Effects of Sexual Offender Treatment on Recidivism: An International Analysis of Sound Quality Evaluations." *Journal of Experimental Criminology* 11 (2015): 597-630.

Schwitzgebel, Eric, Garza Mara. "A Defense of the Rights of Artificial Intelligences." *Midwest Studies in Philosophy* 39, no. 1 (2015): 98-119.

Scruton, Roger. *Sexual Desire*. London: Widenfeld & Nicolson, 1986.

Seibt, Johanna, Hakli Raul, Norskov Marco, eds. *Sociable Robots and the Future of Social Relations: Proceedings of Robo-Philosophy*. Amsterdam: IOS Press, 2014.

Seto, Michael C. *Pedophilia and Sexual Offending Against Children: Theory, Assessment, and Intervention*. Washington, DC: American Psychological Association, 2008.

———. "Is Pedophilia a Sexual Orientation?" *Archives of Sexual Behavior* 41 (2012): 231-236.

———. *Pedophilia and Sexual Offending Against Children: Theory, Assessment, and Intervention* (2nd ed.). Washington, DC: American Psychological Association, 2018.

Seto, Michael C., Cantor James M., Blanchard Ray. "Child Pornography Offenses Are a Valid Diagnostic Indicator of Pedophilia." *Journal of Abnormal Psychology* 115 (2006): 610-615.

Seto, Michael C., Eke Angela W. "The Criminal Histories and Later Offending of Child Pornography Offenders." *Sexual Abuse: A Journal of Research and Treatment* 17 (2005): 201-210.

Seto, Michael C., Lalumière Martin L. "A Brief Screening Scale to Identify Pedophilic Interests Among Child Molesters." *Sexual Abuse: A Journal of Research and Treatment* 13 (2001): 15-25.

Seto, Michael C., Wood J. Michael, Babchishin Kelly M., Flynn Sheri. "Online Solicitation Offenders Are Different From Child Pornography Offenders and Lower Risk Contact Sexual Offenders." *Law and Human Behavior*, 36, no. 4 (2012): 320-330.

Sharkey, Noel, van Wynsberghe Aimee, Robbins Scott, Hancock Eleanor. *Our Sexual Future with Robots. A Foundation for Responsible Robotics Consultation Report*. The Hague, The Netherlands: FRR, 2017.

Sharkey, Amanda Jane, Sharkey Noel. "Granny and the Robots: Ethical Issues in Robot Care for the Elderly." *Ethics and Information Technology* 14, no. 1 (2012): 27-40.

Sharkey, Amanda Jane. "Dignity, Older People, and Robots." *Conference Proceedings 2013*.

Sharkey, Noel. "Saying No to Lethal Autonoumous Targeting." *Journal of Military Ethics* 9, no. 4 (2010): 369-383.

Sharkey, Noel, Sharkey Amanda Jane. "The Crying Shame of Robot Nannies." *Interaction Studies* 11, no. 2 (2010): 161-190.

——. "The Rights and Wrongs of Robot Care." In *Robot Ethics: The Ethical and Social Implications of Robotics*, edited by Patrick Lin, Keith Abney, George A. Bekey, 267-282. Cambridge: MIT Press, 2011.

Sheldon, Kerry, Howitt Dennis. "Sexual Fantasy in Paedophile Offenders: Can Any Model Explain Satisfactorily New Findings from a Study of Internet and Contact Sexual Offenders?" *Legal and Criminological Psychology* 13 (2008): 137-158.

Shelley, Mary. *Frankenstein*. Oxford: Oxford University Press, 2008.

Sherry, John L. "The Effects of Violent Video Games on Aggression. A Meta-Analysis." *Human Communication Research* 27, no. 3 (2001): 409–431.

Sherry, John L. "Violent Video Games and Aggression: Why Can't We Find Links?" In *Mass Media Effects Research: Advances Through Meta-Analysis*, edited by Raymond W. Preiss, Barbara Mae Gayle, Nancy Burrell, Mike Allen, Jennings Bryant, 231-248. Mahwah (N.J.): Erlbaum, 2007.

Silvestri, Pino, "Migliaia di uomini giapponesi hanno una fidanzata virtuale chiamata Rinko." *Virtualblog*, January 24, 2014.

Singer, Peter. "Actual Consequence Utilitarianism." *Mind*, 86 (1977): 67–77.

——. *Practical Ethics*. Cambridge: Cambridge University Press, 2nd edition, 1993.

——. "Heavy Petting." *Nerve* (March/April 2001). www.nerve.com/Opinions/Singer/ heavyPetting.

Sisto, Davide. *La morte si fa social*. Torino: Bollaghi Boringhieri, 2018.

Smith, Adam. *Theory of Moral Sentiments*. Cambridge: Cambridge University Press, 2002.

Soble, Alan. *The Philosophy of Sex: Contemporary Readings, (1a ed.)*. Totowa: Rowman and Littlefield, 1980.

Soble, Alan. *Sexual Investigations*. New York: New York University Press, 1996.
——. *Pornography, Sex and feminism*. Amherst (New York): Prometheus Books, 2002.
Sorell, Tom, Heather Draper. "Robot Carers, Ethics, and Older People." *Ethics Inf Technol* 16 (2014): 183-195.
Sparrow, Robert. "The March of the Robot Dogs." *Ethics and Information Technology* 4, no. 4 (2002): 305-318.
——. "The Turing Triage Test." *Ethics and Information Technology* 6 (2004): 203-213.
——. "Killer Robots." *Journal of Applied Philosophy*, 24, no. 1 (2007): 62-77.
——. "Can Machines Be People? Reflections on the Turing Triage Test." In *Robot Ethics. The Ethical and Social Implications of Robotics*, edited by Patrick Lin, Keith Abney, George A. Bekey, 301-315. Cambridge, Massachusetts: The MIT Press, 2012.
——. "War without Virtue?" In *Killing by Remote Control*, edited by B. J. Strawser, 84-105. Oxford and New York: Oxford University Press, 2013.
——. "Robots in Aged Care: A Dystopian Future?" *AI and Society*, published online November 10 (2015): 445-454.
——. "Robots, Rape, and Representation." *International Journal of Social Robotics* 9, no. 4 (2017): 465-477.
Sparrow, Robert, Howard Mark. *When Human Beings Are Like Drunk Robots: Driverless Vehicles, Ethics, and the Future of Transport*. Transportation Research Part C. Published Online First, May 9, 2017.
Strikwerda, Litska. "Present and Future Instances of Virtual Rape in Light of Three Categories of Legal Philosophical Theories on Rape." *Philos. Technol.* 28 (2015): 491-510.
——. "Legal and Moral Implications of Child Sex Robots." In *Robot Sex. Social and Ethical Implications*, edited by John Danaher, Neil McArthur, 133-151. Cambridge (Massachusetts): The MIT Press, 2017.
Strossen Nadine. *Defending Pornography: Free Speech, Sex, and the Fight for Women's Rights*. New York: Scribner, 1995.
Sullins, John P. "When Is a Robot a Moral Agent?" *International Review of Information Ethics* 6, no. 12 (2006): 23-30.
——. "Robots, Love, and Sex: The Ethics of Building a Love Machine." *IEEE Transactions on Affective Computing* 3, no. 4 (2012): 398-409.

———. "Applied Professional Ethics for the Reluctant Roboticist." In *Proceedings of the Emerging Policy and Ethics of Human-Robot Interaction*, Workshop at HRI, Portland, OR, USA, 2-5 March 2015.

Szczuka, Jessica M., Krämer Nicole C. "Not Only the Lonely – How Men Explicitly and Implicitly Evaluate the Attractiveness of Sex Robots in Comparison to the Attractiveness of Women, and Personal Characteristics Influencing this Evaluation." *Multimodal Technologies and Interact* 1, no. 3 (2017): 1-18.

Szczuka, Jessica M., Krämer Nicole C. *Robotic Love Rival? A Quantitative Study on the Jealousy of Women Evoked by Other Women and Robots*. Conference Paper, Technology, Mind and Society, April 2018.

Szycik, Gregor R., Mohammadi Bahram, Münte Thomas F., Wildt Bert T. te. "Lack of Evidence that Neural Empathic Responses Are Blunted in Excessive Users of Violent Video Games: An fMRI Study." *Frontiers in Psychology* 6 (2017): 174.

Synder, Charles R., Higgins Raymond L. "Excuses: Their Effective Role in the Negotiation of Reality." *Psychological Bulletin* 104 (1988): 23-35.

Tamura, Toshyho, Yonemitsu Satomi, Itoh Akiko, Oikawa Daisuke, Kawakami Akiko, Hogashi Yuji, Fujimooto Toshiro, Nalajima Kazuki. "Is An Entertainment Robot Useful in the Care of Elderly People with Severe Dementia?" *J Gerontol A Biol Sci Med Sci.* 59, no. 1 (2004): 83-5.

Taormino, Tristan. *The Ultimate Guide to Anal Sex for Women*. San Francisco: Cleis Press, 1998.

Taylor, Shelley E., Brown Jonathon D. "Illusion and Well-Being: A Social Psychological Perspective on Mental Health." *Psychological Bulletin* 103 (1988): 193-210.

Tegmark, Max. *Life 3.0: Being Human in the Age of Artificial Intelligence*. New York: Knopf, 2017.

Thomsen, Frej Klem. "Prostitution, Disability and Prohibition." *Journal of Medical Ethics* 41, no. 6 (2015): 451-9.

Ticknor, Bobbie, Tillinghast Sherry. "Virtual Reality and the Criminal Justice System: New Possibilities for Research, Training, and Rehabilitation." *Journal of Virtual Worlds of Research* 4 (2011): 3-44.

Tiefer Leonore. *Sex Is Not a Natural Act*. Boulder (Colorado): Westview Press, 1995.

Tolstoy, Lev, *Anna Karenina*. Translated by Richard Pevear and Larissa Volokhonsky, London: Penguin Classics, 2004.

Torjesen, Ingrid. "Society Must Consider Risks of Sexbots." *The British Medical Journal* 358 (2017).
Torresen, Jim. "A Review of Future and Ethical Perspectives of Robotics and AI." *Frontiers in Robotics and AI*, 4, no. 75 (2018): 1-10.
Tronto, Joan C., *Moral Boundaries: A Political Argument for an Ethic of Care*. New York: Routledge, 1993.
Turkle, Sherry. *The Second Self: Computers and the Human Spirit*. New York: Simon and Schuster, 1984.
———. "Authenticity in the Age of Digital Companions." *Interaction Studies* 8 (2007), 501-517.
Ulivieri, Max. "Il sesso per un disabile? Diritto ed esigenza". Interviewed by Cecilia Pierami. *Tgcom24*, February 2, 2013. www.tgcom24.mediaset.it/cronaca/articoli/1080674/il-sesso-per-un-disabile-diritto-ed-esigenza.shtml
———, ed. *Loveability. L'assistenza sessuale per le persone con disabilità*. Trento: Erickson, 2014.
Vaccari, Alessio. *Le etiche della virtù*. Firenze: Le Lettere, 2012.
Vallor, Shannon. "Carebots and Caregivers: Sustaining the Ethical Ideal of Care in the 21st Century." *Philosophy and Technology* 24, no. 3 (2011): 251-268.
Verso, Francesco. "Futuro di mezza estate." In Maurizio Balistreri, Francesco Verso, *I non nascituri. Future Fiction* (5-31). Roma: Future Fiction, 2018.
———. *E-Doll*. Mondadori: Milano, 2009.
Wakeling, Helen C., Mann Ruth E., Carter Adam J. "Do Low-Risk Sexual Offenders Need Treatment?" *The Howard Journal* 51 (2012): 286-299.
Wallach, Wendell, Allen Colin. *Moral Machines. Teaching Robots Right from Wrong*. Oxford: Oxford University Press, 2009.
Walsh, Toby. "Will Robots Bring about the End of Work?" *The Guardian*, October 1, 2017.
Ward, Tony, Hudson Stephen M., Marshall William L. "Cognitive Distortions in Sex Offenders: An Integrative Review." *Clinical Psychology Review* 17 (1997): 479-507.
Ward, Tony, Hudson Stephen M. "Sexual Offenders' Implicit Planning: A Conceptual Model." *Sexual Abuse: A Journal of Research and Treatment*, 12 (2000): 189-202.
Way, Ben. *Jobacalypse: The End of Human Jobs and How Robots Will Replace Them*. CreateSpace Indipendent Publishing Platform, 2013.

Webb, Liane, Craissati Jackie, Keen Sarah. "Characteristics of Internet Child Pornography Offenders: A Comparison with Child Molesters." *Sexual Abuse* 19 (2007): 449-465.

Webber, Jonathan. "Sex." *Philosophy* 84, no. 2 (2009): 233-250.

Weitzer, Ronald. "New Directions in Research on Prostitution." *Crime, Law and Social Change* 43, no. 4-5 (2005): 211-235.

Wennerscheid, Sophie. "Posthuman Desire in Robots and Science Fiction." In *Love and Sex with Robots. Third International Conference LSR 2017*, edited by Adrian David Cheok, David Levy, 37-50. Cham (Switzerland): Springer, 2018.

——. *Sex Machina. Zur Zukunft des Begehrens*. Berlin: Matthes & Seitz Berlin, 2019.

Westin, Anna. "The Harms of Prostitution: Critiquing Moen's Argument of No Harm." *Journal of Medical Ethics* 40, no. 2 (2013): 86-87.

Whipple, Beverly. "The Benefits of Sexual Expression on Physical Health." *Sexologies* 17, Supplement 1 (2008): 545–546.

Whitby, Blay. "Do You Want a Robot Lover? The Ethics of Caring Technologies." In *Robot Ethics: The Ethical and Social Implications of Robotics*, edited by Patrick Lin, Keith Abney, George A. Bekey 233-248. Cambridge: MIT Press, 2011.

Wilcox, Daniel T., Garrett Tanya, Harkins Leigh, eds. *Sex offender Treatment. A Case Study Approach to Issues and Interventions*. Chichester, UK: Wiley-Blackwell, 2015.

Wilks, Yorick. *Artificial Intelligence. Modern Magic or Dangerous Future?* London, Uk: Icon Books Ltd, 2019.

Wilson, David, Jones Timothy. "'In My Own World': A Case Study of a Paedophile's Thinking and Doing and His Use of the Internet." *The Howard Journal of Criminal Justice* 47 (2008): 107-120.

Wilson, Glenn D.. *The Secrets of Sexual Fantasy*. London: Dent, 1978.

——. "Gender Differences in Sexual Fantasy: An Evolutionary Analysis." *Personality and Individual Differences* 22 (1997): 27-31.

Wilson, Robin J.. "Emotional Congruence in Sexual Offenders Against Children." *Sexual Abuse: Journal of Research and Treatment* 11 (1999): 33-47.

Wiseman, Eva. "Sex, Love and Robots: Is This the End of Intimacy?" *The Guardian* (Online), December 13, 2015.

Woodhouse, Susan S., Gelso Charles J. "Volunteer Client Adult Attachment, Memory for Insession Emotion, and Mood Awareness:

An Affect Regulation Perspective." *Journal of Counseling Psychology* 55 (2008): 197-208.

Wosk, Julie. *My Fair Ladies. Female Robots, Androids and Other Artificial Eves.* New Brunswick (NJ): Rutgers University Press, 2015.

Wright, Paul J., Tokunaga Robert, Kraus Ashley. "A Meta-Analysis of Pornography Consumption and Actual Acts of Sexual Aggression in General Population Studies." *J Commun* 66, no. 1 (2016): 183-205.

Wripple, Beverly. "The Benefits of Sexual Expression on Physical Health." *Sexologies*, 17, Supplement 1 (2008): 545-546.

Yeoman, Ian, Mars Michelle. "Robots, Men and Sex Tourism." *Futures* 44 (2012): 365-371.

YouGov. *1 in 4 Men Would Consider Having Sex with a Robot*, September 27, 2017. https://today.yougov.com/news/2017/10/02/1-4-men-would-consider-having-sex-robot/

Young, Garry. *Resolving the Gamer's Dilemma.* Cham: Palgrave, 2016.

Zagal, Jose P.. *The Videogame Ethics Reader.* San Diego (CA): Cognella, 2012.

Zara, Georgia. "Il ruolo del diniego nella violenza sessuale: la valutazione dei bisogni criminogenici e dei bisogni di rispondenza." *Gruppi*, 1 (2016): 123-140.

——. *Valutare il rischio in ambito criminologico. Procedure e strumenti per l'assessment psicologico.* Bologna: Il Mulino, 2016.

——. *Il diniego nei sex offender: dalla valutazione al trattamento.* Milano: Raffaello Cortina, 2018.

——. "L'impatto del diniego tra rischio di recidiva e trattamento dei reati sessuali: il beneficio paradossale." *Giornale Italiano di Psicologia*, XLV, no. 2 (2018): 333–360.

Zara, Georgia, Farrington David P. *Criminal Recidivism: Explanation, Prediction and Prevention.* Abingdon (UK): Routledge, 2016.

Zhou, Yuefang, Fischer Martin H., eds. *AI Love You. Developments in Human-Robot Intimate Relationships.* Cham: Springer, 2019.

Zubrycki, Igor, Granosik Grzegorz. "Understanding Therapists' Needs and Attitudes Towards Robotic Support. The Roboterapia Project." *International Journal of Social Robotics* 8 (2016): 553-563.

INDEX

A

Abuse, 33, 38–42, 73, 101, 108, 112–113, 115–116, 119, 123, 126, 129–130, 132, 134, 136, 140–141
Abyss Creations, 22
Addiction, 103, 114, 121
Adultery, 28
Affair, 1, 54, 96, 124, 127
AI, 28, 112, 116, 118, 138, 140, 142
Albinati, Edoardo, 111
Anal sex, 30, 139
Animals, 42, 44, 65, 84, 94–95, 104, 117–118, 120, 131
Anna Karenina, 47, 139
Annie and I (film), 25
Anonymous sex, 58
Aquinas, Thomas, 28
Artificial intelligence (AI), 10, 13, 14, 16, 23, 69, 81, 89, 92, 93
Appel, Jacob M., 111
Asimov, Isaac, 112
Assault, 37, 52–53, 55, 122, 131
Atkinson, Robert D., 112
Authenticity, 65, 72, 140
Autoeroticism, 25–28, 48
Automaton, 15, 108
Autonomous Machines, 8, 15–16
Autonomy, 54

B

Balistreri, Maurizio, 112
Banks, Marian R., 112
Barbero, Carola, 113
Barkan, Howard, 64
Beard, Jack, 113
Beaver, Kevin M., 56
Bemelmans, Roger, 113–114
Bendel, Oliver, 114
Benefits, 11–12, 32, 47–49, 66, 72–73, 115–116, 141–142
Bennett, Casey Carroll, 114
Bernstein, Elizabeth, 114
Beschorner, Thomas, 114
Bestiality, 28, 104, 119
Betrayal, 77, 81–84, 89
Blackford, Russell, 115
Bostrom, Nick, 115
Brodbeck, Luzius, 115
Brody, Stuart, 32

Brown, Rick, 116
Bryson, Joanna, 116
Bullock, Caroline, 116
Burnet, Dean, 12
Buss, David M., 25

C

Carpenter, Julie, 117
Carr, Nicholas, 117
Casalini, Brunella, 117
Casetta, Elena, 117
Casini, Stefano, 117
Casual sex, 58
Cheok, Adrian David, 117
Child Robot, 108
Coeckelbergh, Mark, 117
Cohen-Green, Cheryl, 9
Commercial relationships, 65
Companionship, 8, 42, 48, 90
Consciousness, 107
Consent, 53, 63, 111, 123–125
Couric, Katie, 118
Cox-George, Chantal, 118
Crawford, Neta C., 118

D

Damiano, Luisa, 89, 97
Danaher, John, 118–119
De Boer, Tracy, 119
Dehumanisation, 101
Dekkers, Midas, 119
Devlin, Amanda, 120
Di Nucci, Ezio, 120
Diamond, Milton, 120
Dignity, 35, 54, 64, 137
Dildo, 1, 81
Disease, 17, 63, 67–68

Döring, Nicola, 120
Duggan, Lisa, 120
Dumouchel, Paul, 120
Dworkin, Andrea, 121

E

Earp, Brian D., 121
Easton, Dossie, 121
Education, 12, 66, 131
Ejaculation, 105
Elliott, Larry, 121
Embodied, 14
Empathy, 56, 63, 65, 96, 118–119, 127
Eskens, Romy, 122
Exchange, 65–66, 78, 80, 87
Ex Machina (film), 18, 85

F

Fantasy, 2–3, 23, 26, 34, 36, 40, 44, 47, 51, 55, 57, 62–63, 68, 78, 81–82, 84, 96, 104, 106, 108, 112, 114–115, 117, 123–124, 126–128, 137, 141
Farley, Melissa, 122
Faustino, Maria João, 122
Ferguson, Christopher J., 122–123
Fetish, 3
Fidelity, 5
Flaubert, Gustave, 88
Forsch-Villaronga, Eduard, 70
Ford, Martin, 123
Frankenstein, 14, 137
Freedom, 35, 54, 69
Friendship, 44, 119

G

Garofalo Geymonat, Giulia, 123
Gender, 19, 43, 58, 61, 78, 106, 127, 141
Gentile, Douglas, 124
George, Robert P., 124
Goethe, Johann Wolfgang, 88
Goldman, Alan H., 124
Gomes, Leonardo M., 124
Goodal, Noah Joseph, 124
Google, 12
Group sex, 58
Gunkel, David J., 124
Gutiu, Sinziana, 124

H

Haldane, Andy, 9
Happiness, 32, 45
Harari, Yuval Noah, 125
Harassment, 17, 33, 107
Harm, 14, 16, 19, 28–29, 35, 38–39, 52, 54, 64, 104, 107, 121, 131, 141
Harmony, 106
Hauskeller, Michael, 125
Health, 12, 32, 45, 64, 72, 74–75, 100, 108, 114–115, 118, 122–123, 126, 135, 139, 141–142
Her (film), 18
Heterosexual, 23, 30, 81, 106, 126
Hines, Douglas, 21
Homosexuality, 28
Huesmann, L. Rowell, 126

Humanoid robot, 18, 81, 91, 97–101, 116
Humans (TV series), 85
Hume, David, 126

I

Imagination, 29, 59, 83–84, 95–97, 105, 108, 114–115
Incest, 28
Inequality, 107
Infidelity, 67, 78, 80–82, 135
Integrity, 84, 95
Intelligence, 10, 13–14, 16, 23, 69, 81, 89, 92–93, 97, 112–113, 115, 117–118, 123, 136, 139, 141
Intimacy, 2, 25–26, 41, 62, 65, 105, 114, 129, 135, 141
I Robot (film), 14

K

Kanamori, Masao, 126
Kilburn, John, 56
Killer robots, 54,
Kathleen Richardson, 53
Killing, 16, 36–38, 54, 57, 108, 116, 118, 138
Krämer, Nicole C., 23, 81
Krause, Florian, 90
Kühn, Simone, 127

L

Lake, Emma, 22, 68
Langton, Rae, 127
Lars and the Real Girl (film), 85

Index

Lecaldano, Eugenio, 127
Levy, David, 127
Lewin, Ben, 70
Liberati, Nicola, 128
Lin, Patrick, 128
Love, 2, 9, 18–19, 23, 25–28, 31, 40, 43–46, 48, 52, 54–55, 60, 65, 75, 78–79, 82–94, 96, 99, 101–105, 108, 113–115, 117, 119, 121, 124, 127–128, 132, 135, 138–139, 141–142
LovePlus, 90
Lovotics, 85, 117, 128
Luck, Morgan, 128
Lunceford, Brett, 129

M

MacKinnon, Catharine, 57
Markey, Patrick M., 129
Marriage, 26, 58, 65, 82, 111
Marrone, Pierpaolo, 129
Masturbation, 16, 25–26, 28–29, 32, 64, 71, 80
McArthur, Neil, 130
McAnulty, Richard, 64
McCormick, Matt, 34, 37
McDorman, Karl F., 130
McElroy, Wendy, 130
McEwan, Ian, 130
Meston, Cindy M., 130
Micklethwaite, Jamie, 130
Midgley, Mary, 130
Mistreatment, 51
Moen, Ole Martin, 130–131
Monitoring, 11
Monto, Martin A., 131
Moran, Rachel, 131

Morgan, Seiriol, 131
Morgan, William, 131
Mori, Masahiro, 131
My Sex Robot (documentary), 30

N

Nicolaci da Costa, Pedro, 132
Novelty, 13, 49
Nussbaum, Martha, 132
Nyholm, Sven, 132

O

Objectification, 63, 105
Oral sex, 30
Orgasm, 16, 19, 27–28, 53, 65, 71, 78, 105–106
Owsianik, Jenna, 132

P

Pain, 1, 35, 37, 40, 94, 127
Palazzani, Laura, 132
Parfit, Derek, 132
Patridge, Stephanie, 34
Pedophilia, 37, 55, 128, 131, 136
Penetration, 19, 70, 105–106
Penis, 30, 59, 105–106
Person, 3, 8–13, 15–19, 22–23, 25–49, 51–52, 54–75, 77–108, 121, 124, 128, 134, 137–139
Perversion, 25, 108
Pöschl, Sandra, 74
Plato, 102, 133
Pleasure, 1, 8, 11, 15–16, 18–19, 23–25, 27–37, 41, 47–48, 51–52, 54, 56, 58–61, 64–65,

68–69, 72, 74–75, 79–80, 82, 94, 104–106, 122
Pornography, 26, 35, 52, 57–58, 68, 114, 119–122, 126, 130–132, 136, 138, 141–142
Predictions, 9, 68, 93
Preferences, 9, 22, 27, 30, 61, 69, 86, 91, 106
Pregnancy, 61
Primoratz, Igor, 133
Privacy, 55
Prostitution, 17–18, 32, 63–64, 66–70, 72–74, 101, 108, 116, 121–122, 130–131, 134–135, 139, 141

Q

Quattrini, Fabrizio, 133

R

Rachels, James, 133
Rape, 72, 120, 138
RealDoll, 22
Reciprocity, 65
Relationships, 7, 15–19, 22–23, 26–31, 40–44, 46, 48, 52, 54, 58, 62, 65, 74, 84, 90, 92–93, 100–101, 103–104, 113–114, 121, 128, 142
Representations, 35–36, 39
Reproduction, 26–28, 105, 109
Reproductive, 105, 118
Responsibility, 10–11, 17, 28, 61, 90, 113, 118, 126
Ricca, Jacopo, 134
Richardson, Kathleen, 134
Rights, 9, 35–36, 82, 96, 98, 101, 108, 111, 118, 120–121, 124, 134, 136–138
Rigotti, Carlotta, 134
Risks, 9, 11, 16, 32, 64, 68, 123, 140
Robinson, Hayley, 134
Rocky, 21
Rodogno, Raffaele, 134
Romantic love, 87
Roxxxy, 21–22, 33, 128
Royakkers Lamber, 135

S

Sanders, Teela, 135
Satisfaction, 25, 90
Scruton, Roger, 136
Seibt, Johanna, 136
Self-determination, 85
Sex dolls, 60, 68, 116, 120, 129, 134
Sex work, 2, 68, 74, 118, 135
Sex workers, 2, 117, 128, 135
Sexual assault, 122, 131
Sexual Orientation, 22, 106, 136
Sexual stimulation, 41
Sexually transmitted diseases, 17
Sharkey, Amanda Jane, 137
Sharkey, Noel, 136–137
Sherry, John L., 137
Silvestri, Pino, 137
Singer, Peter, 137
Sisto, Davide, 137
Slavery, 14, 66, 98
Slaves, 98, 116

Smith, Adam, 137
Soble, Alan, 137–138
Sodomy, 30
Sparrow, Robert, 138
Stereotype, 106, 124
Stigmatization, 64
Strikwerda, Litska, 138
Subjectivity, 64
Submission, 35–36
Sullins, John, 138
Superintelligence, 14, 16, 115
Star Trek: The Next Generation (TV series), 61
Status (moral), 36
Synder, Charles R., 139
Synthea Amatus, 22, 33
Szczuka, Jessica M., 139
Szycik, Gregor R., 139

T

Tamagotchi, 89–90
Tamura, Toshyho, 139
Taormino, Tristan, 139
Taylor, Shelley E., 139
Technology, 1–2, 6, 9, 14, 18, 21, 32–33, 56, 61, 77–79, 86–87, 89–90, 93–94, 105–107, 115, 118, 121, 123, 127–128, 130–131, 133–134, 137–141
Teledildonics, 41, 80, 106
Terminator (film), 14,
Therapist, 72
The Session (film), 70
Thomsen, Frej Klem, 139
Training, 8, 13, 71, 139
Transgender, 106
Trauma, 32, 115

Tronto, Joan, 140
True Companion, 21, 33
Turing test, 18, 108
Turkle, Sherry, 140

U

Ulivieri, Max, 140
Uncanny valley, 91, 127, 129–131

V

Vaccari, Alessio, 140
Vagina, 28, 30, 59, 105–107
Verso, Francesco, 140
Vibrator, 28, 79, 81, 107
Victims, 17, 37, 52, 54
Violence, 2, 14, 16–17, 27–29, 33–38, 40–42, 48, 51–75, 89, 92, 96, 101, 108–109, 115, 122–125, 129, 133
Virtual reality, 1, 26, 37–39, 41, 89, 114, 134–135, 139
Virtue, 2–3, 29, 34, 120, 138
Voyeuristic sex, 58

W

Wall-E (film), 18
Walsh, Toby, 140
Way, Ben, 140
Webber, Jonathan, 141
Weitzer, Ronald, 141
Well-being, 29, 32, 35, 48, 72, 100, 139
Westin, Anna, 141
Westworld (TV series), 96
Whipple, Beverly, 141

Whitby, Blay, 141
Woody, Allen, 25, 55
Wosk, Julie, 142
Wright, Paul J., 142

Y

Young, Garry, 142

Z

Zagal, Jose P., 142